Higher

Physics

Specimen Question Paper
(based on the 2000 SCE Higher Paper)

2000 Exam

2001 Exam

2002 Exam

2003 Exam

© Scottish Qualifications Authority

All rights reserved. Copying prohibited. No part of this publication may be reproduced, stored in a retrieval system, or transmitted in any form or by any means, electronic, mechanical, photocopying, recording or otherwise.

First exam published in 2000.
Published by
Leckie & Leckie, 8 Whitehill Terrace, St. Andrews, Scotland KY16 8RN
tel: 01334 475656 fax: 01334 477392
enquiries@leckieandleckie.co.uk www.leckieandleckie.co.uk

Leckie & Leckie Project Team: Peter Dennis; John MacPherson; Bruce Ryan; Andrea Smith

ISBN 1-84372-129-5

A CIP Catalogue record for this book is available from the British Library.

Printed in Scotland by Scotprint.

Leckie & Leckie is a division of Granada Learning Limited, part of Granada plc.

Introduction

Dear Student,

This past paper book provides you with the perfect opportunity to put into practice everything you should know in order to excel in your exams. The compilation of papers will expose you to an extensive range of questions and will provide you with a clear idea of what to expect in your own exam this summer.

The past papers represent an integral part of your revision: the questions test not only your subject knowledge and understanding but also the examinable skills that you should have acquired and developed throughout your course. The answer booklet at the back of the book will allow you to monitor your ability, see exactly what an examiner looks for to award full marks and will also enable you to identify areas for more concentrated revision. Make use too of the tips for revision and sitting your exam to ensure you perform to the best of your ability on the day.

Practice makes perfect. This book should prove an invaluable revision aid and will help you prepare to succeed.

Good luck!

Acknowledgements

Every effort has been made to trace the copyright holders and to obtain their permission for the use of copyright material. Leckie & Leckie will gladly receive information enabling them to rectify any error or omission in subsequent editions.

HIGHER SPECIMEN QUESTION PAPER

X069/301

NATIONAL QUALIFICATIONS

Time: 2 hours 30 minutes

PHYSICS HIGHER

Specimen Question Paper (based on the 2000 SCE Higher paper)

Read Carefully

1. All questions should be attempted.

Section A (questions 1 to 20)

2. Check that the answer sheet is for Physics Higher (Section A).
3. Answer the questions numbered 1 to 20 on the answer sheet provided.
4. Fill in the details required on the answer sheet.
5. Rough working, if required, should be done only on this question paper, or on the first two pages of the answer book provided—**not** on the answer sheet.
6. For each of the questions 1 to 20 there is only **one** correct answer and each is worth 1 mark.
7. Instructions as to how to record your answers to questions 1–20 are given on page three.

Section B (questions 21 to 29)

8. Answer questions numbered 21 to 29 in the answer book provided.
9. Fill in the details on the front of the answer book.
10. Enter the question number clearly in the margin of the answer book beside each of your answers to questions 21 to 29.
11. Care should be taken to give an appropriate number of significant figures in the final answers to calculations.

DATA SHEET
COMMON PHYSICAL QUANTITIES

Quantity	Symbol	Value	Quantity	Symbol	Value
Speed of light in vacuum	c	3.00×10^8 m s^{-1}	Mass of electron	m_e	9.11×10^{-31} kg
Magnitude of the charge on an electron	e	1.60×10^{-19} C	Mass of neutron	m_n	1.675×10^{-27} kg
Gravitational acceleration	g	9.8 m s^{-2}	Mass of proton	m_p	1.673×10^{-27} kg
Planck's constant	h	6.63×10^{-34} J s			

REFRACTIVE INDICES
The refractive indices refer to sodium light of wavelength 589 nm and to substances at a temperature of 273 K.

Substance	Refractive index	Substance	Refractive index
Diamond	2·42	Water	1·33
Crown glass	1·50	Air	1·00

SPECTRAL LINES

Element	Wavelength/nm	Colour	Element	Wavelength/nm	Colour
Hydrogen	656	Red	Cadmium	644	Red
	486	Blue-green		509	Green
	434	Blue-violet		480	Blue
	410	Violet			
	397	Ultraviolet			
	389	Ultraviolet			

Lasers

Element	Wavelength/nm	Colour
Carbon dioxide	9550, 10590	Infrared
Helium-neon	633	Red

Element	Wavelength/nm	Colour
Sodium	589	Yellow

PROPERTIES OF SELECTED MATERIALS

Substance	Density/ kg m^{-3}	Melting Point/ K	Boiling Point/ K
Aluminium	2.70×10^3	933	2623
Copper	8.96×10^3	1357	2853
Ice	9.20×10^2	273	....
Sea Water	1.02×10^3	264	377
Water	1.00×10^3	273	373
Air	1·29	....	....
Hydrogen	9.0×10^{-2}	14	20

The gas densities refer to a temperature of 273 K and a pressure of 1.01×10^5 Pa.

SECTION A

For questions 1 to 20 in this section of the paper, an answer is recorded on the answer sheet by indicating the choice A, B, C, D or E by a stroke made in ink in the appropriate box of the answer sheet—see the example below.

EXAMPLE

The energy unit measured by the electricity meter in your home is the

    A ampere

    B kilowatt-hour

    C watt

    D coulomb

    E volt.

The correct answer to the question is B—kilowatt-hour. Record your answer by drawing a heavy vertical line joining the two dots in the appropriate box on your answer sheet in the column of boxes headed B. The entry on your answer sheet would now look like this:

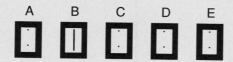

If after you have recorded your answer you decide that you have made an error and wish to make a change, you should cancel the original answer and put a vertical stroke in the box you now consider to be correct. Thus, if you want to change an answer D to an answer B, your answer sheet would look like this:

If you want to change back to an answer which has already been scored out, you should enter a tick (✓) to the RIGHT of the box of your choice, thus:

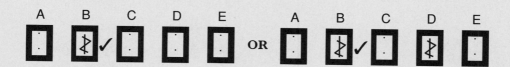

SECTION A

Answer questions 1 to 20 on the answer sheet.

1. A speed skier crosses the **start-line** of a straight 200 metre downhill course with a speed of $30\,\text{m s}^{-1}$. She accelerates uniformly all the way down and takes 5 s to cover the course.

   What is her speed as she crosses the **finish-line**?

   A $\ \ 30\,\text{m s}^{-1}$
   B $\ \ 35\,\text{m s}^{-1}$
   C $\ \ 40\,\text{m s}^{-1}$
   D $\ \ 45\,\text{m s}^{-1}$
   E $\ \ 50\,\text{m s}^{-1}$

2. While being towed behind a speedboat, John and his paraglider are moving horizontally with a constant velocity, as shown below.

   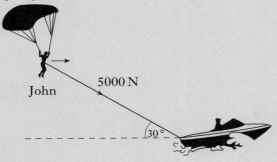

   The towing cable exerts a force of 5000 N on John and his paraglider. What is the horizontal resistive force acting on John and his paraglider?

   A $\ \ 2500\,\text{N}$
   B $\ \ 4330\,\text{N}$
   C $\ \ 5774\,\text{N}$
   D $\ \ 7500\,\text{N}$
   E $\ \ 10\,000\,\text{N}$

3. The diagram below shows the masses and velocities of two trolleys just before they collide on a level bench.

   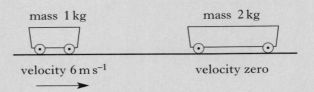

   After the collision, the two trolleys move along the bench joined together.

   How much kinetic energy is lost as a result of this collision?

   A $\ \ 0\,\text{J}$
   B $\ \ 6\,\text{J}$
   C $\ \ 12\,\text{J}$
   D $\ \ 18\,\text{J}$
   E $\ \ 24\,\text{J}$

4. Water flows over a waterfall, which is 120 m high, at a rate of $1 \times 10^6 \,\text{kg s}^{-1}$.

   Assuming that the acceleration due to gravity is $9.8\,\text{m s}^{-2}$, which of the following gives the power delivered by the water in falling?

   A $\ \ 8.5 \times 10^2\,\text{W}$
   B $\ \ 8.3 \times 10^3\,\text{W}$
   C $\ \ 1.2 \times 10^7\,\text{W}$
   D $\ \ 1.2 \times 10^8\,\text{W}$
   E $\ \ 1.2 \times 10^9\,\text{W}$

5. Which of the following statements about the absolute zero of temperature is/are true?

   I The absolute zero of temperature is $-273\,\text{K}$.

   II At absolute zero, movement of molecules ceases in an ideal gas.

   III At absolute zero, the mass of an ideal gas is zero.

   A I only
   B II only
   C I and II only
   D II and III only
   E I, II and III

6. The weight of a solid object is 50 N and its volume is 1×10^{-3} m^3. Assuming that the gravitational field strength is 9.8 N kg^{-1}, the density of this object is

A 1.0×10^{-4} kg m^{-3}

B 5.0×10^{-2} kg m^{-3}

C 5.1×10^{3} kg m^{-3}

D 5.0×10^{4} kg m^{-3}

E 4.9×10^{5} kg m^{-3}.

7. Which of the following gives the correct relationship between the pressure and temperature of a fixed mass of an ideal gas at constant volume?

A The pressure is directly proportional to the temperature in °C.

B The pressure is inversely proportional to the temperature in °C.

C The pressure is directly proportional to the temperature in K.

D The pressure is inversely proportional to the temperature in K.

E The expression (pressure × temperature in K) is constant.

8. In an oscilloscope tube, an electron is accelerated through a potential difference of 15 000 V. The magnitude of the charge on the electron is 1.6×10^{-19} C.

The gain in kinetic energy of the electron, in joules, is given by

A $1.6 \times 10^{-19} \times 15\,000$

B $\dfrac{1.6 \times 10^{-19}}{15\,000}$

C $\dfrac{15\,000}{1.6 \times 10^{-19}}$

D $0.5 \times 1.6 \times 10^{-19} \times (15\,000)^2$

E $0.5 \times 15\,000 \times (1.6 \times 10^{-19})^2$.

9. A resistor of resistance 100 Ω is rated at 4 W. What is the maximum voltage which can be applied across the resistor without exceeding its power rating?

A 0·04 V

B 5 V

C 20 V

D 25 V

E 400 V

10. The Wheatstone bridge circuit below includes a light dependent resistor (LDR). The resistance of the LDR decreases as the light intensity increases.

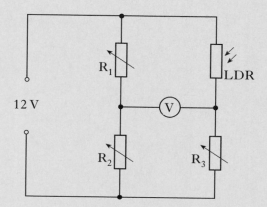

The resistance of R_1 is adjusted so that the reading on the voltmeter is 0 V. The LDR is now illuminated with a bright lamp and the voltmeter reading changes.

Which of the following actions would return the voltmeter reading to zero?

 I Increasing the value of R_3

 II Decreasing the value of R_1

 III Increasing the value of R_2

A I only

B II only

C III only

D I and III only

E II and III only

11. In the following Wheatstone bridge circuit, the reading on the voltmeter is zero when the resistance R_v of the variable resistor is 1 kΩ.

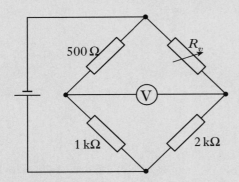

Which of the following would best represent the shape of a graph of the voltmeter reading V against the resistance R_v as it is varied between 990 Ω and 1010 Ω?

A
B
C
D
E

12. The following circuit shows an a.c. supply of constant r.m.s. voltage connected to a resistor and a capacitor in parallel.

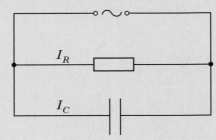

Which pair of graphs shows how the r.m.s. currents I_R and I_C vary as the frequency f of the output of the supply is increased slowly?

A
B
C
D
E

13. The operational amplifier circuit shown below has a 1·5 V d.c. signal at its input.

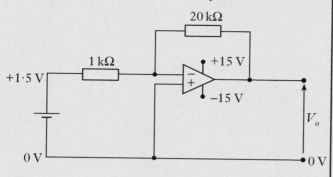

The operational amplifier is operated from a ±15 V supply. Which one of the following gives the most likely value for the output voltage V_o?

A +30 V
B −30 V
C +13 V
D −13 V
E +15 V

14. Microwaves of wavelength 2·8 cm are allowed to pass through the slits S_1 and S_2 in a metal plate.

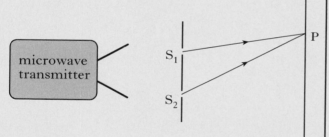

A detector locates a maximum intensity at position P. The path difference $(S_2 P - S_1 P)$ is 16·8 cm.

Which of the following statements about the radiations from S_1 and S_2 is/are true?

I They are coherent.
II They combine destructively at P.
III Their path difference is a whole number of wavelengths.

A II only
B III only
C I and III only
D II and III only
E I, II and III

15. The diagram below shows a ray of light, of wavelength $6·30 \times 10^{-7}$ m in air, entering a glass block of refractive index 1·50.

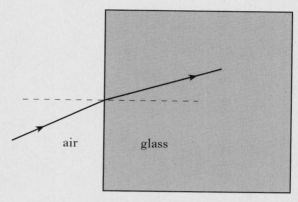

The frequency of the light in the glass block is

A $4·76 \times 10^{14}$ Hz
B $7·14 \times 10^{14}$ Hz
C $1·89 \times 10^{2}$ Hz
D $1·26 \times 10^{2}$ Hz
E $2·10 \times 10^{-15}$ Hz.

16. A ray of light passes from air into substance X, as shown below.

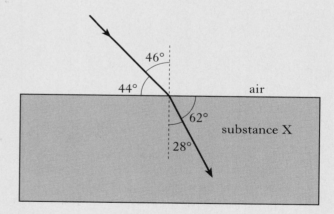

The critical angle for substance X for this light is

A 32·1°
B 40·7°
C 45·0°
D 51·9°
E 90·0°.

17. The intensity of light on a table surface is $6{\cdot}63\,\text{W m}^{-2}$.

The wavelength of the light is 600 nm.

How many photons are incident on each square metre of the table surface every second?

(Planck's constant $h = 6{\cdot}63 \times 10^{-34}\,\text{J s}$)

- A $\quad 2{\cdot}0 \times 10^{19}$
- B $\quad 6{\cdot}0 \times 10^{27}$
- C $\quad 1{\cdot}0 \times 10^{34}$
- D $\quad 1{\cdot}7 \times 10^{40}$
- E $\quad 5{\cdot}0 \times 10^{48}$

18. Part of the energy level diagram for an atom is shown below.

The magnitude of the energy change from E_1 to E_2 is considerably more than the magnitude of the energy change from E_3 to E_2.

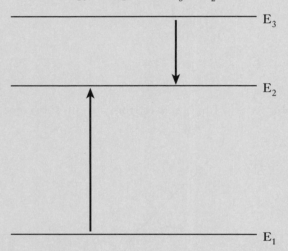

The electron transition from E_3 to E_2 corresponds to a green line in the emission spectrum of the atom.

The electron transition from E_1 to E_2 could correspond to the

- A absorption of green light
- B emission of green light
- C emission of blue light
- D absorption of blue light
- E absorption of red light.

19. Which of the following units is used for measuring **dose equivalent**?

- A Bq
- B Sv
- C Gy
- D Sv s^{-1}
- E Gy s^{-1}

20. Increasing thicknesses of lead are placed between a detector and a gamma source.

Which graph shows how the intensity of the radiation at the detector changes with the thickness of the lead absorber?

A
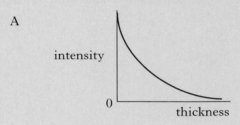

B
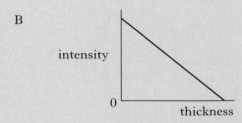

C
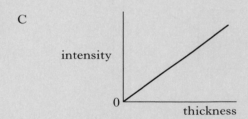

D
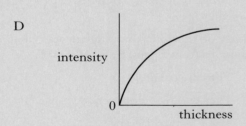

E
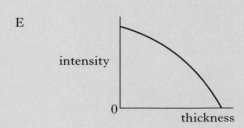

SECTION B

Write your answers to questions 21 to 29 in the answer book.

21. A rocket is used to launch a spacecraft.

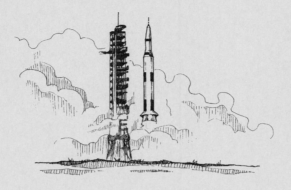

(a) The rocket and spacecraft have a combined mass of 3.0×10^5 kg.

On lift-off from the Earth, the rocket motors produce a force of 3.6×10^6 N.

(i) Draw a diagram of the rocket to show the forces acting on the rocket just after lift-off.

You must name each force.

(ii) Calculate the initial acceleration of the rocket.

(iii) Although the force exerted by the rocket motors remains constant, the acceleration of the rocket increases as the rocket rises.

Give **one** reason why this happens.

(b) A spacecraft has to dock with a module in deep space where the effects of gravity may be considered negligible.

The module should be considered to be at rest. The spacecraft approaches the module with a constant velocity of 2.0 m s^{-1} as shown in the diagram below.

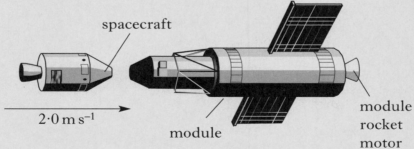

The spacecraft docks with the module.

The mass of the spacecraft is 5.0×10^4 kg and the mass of the module is 1.5×10^5 kg.

(i) Show by calculation that the speed of the spacecraft and module combination is 0.50 m s^{-1}.

(ii) The module fires its motor for 10.0 s and this reduces the speed of the combination to zero.

Calculate the size of the average force exerted by the motor.

Marks

4

4

(8)

22. A radio controlled aircraft is flown over a course from A to B and then to C as shown below.

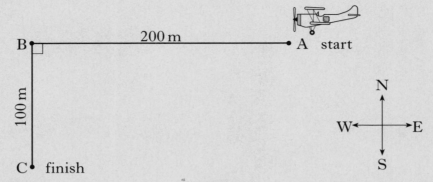

(a) During one run there is no wind and the aircraft is flown at a constant speed of $25\,\text{m s}^{-1}$.

   (i) Calculate the time the aircraft takes to complete the course.

   (ii) When it reaches C, what is the resultant displacement of the aircraft relative to its starting position?   **3**

22. (continued)

(b) A tennis player strikes a ball with his racket just as the ball reaches the ground.

The ball leaves the racket with a speed of $6.0 \, \text{m s}^{-1}$ at $50°$ to the ground as shown in the diagram below. The effect of air resistance should be ignored.

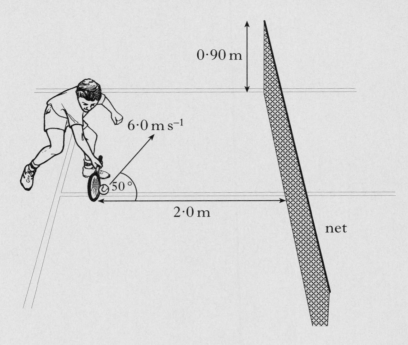

(i) Calculate the vertical component of the initial velocity of the ball.

(ii) Calculate the horizontal component of the initial velocity of the ball.　　**2**

(c) As shown in the diagram, when the ball is struck, it is $2.0 \, \text{m}$ from the base of the net.

(i) Calculate the time taken for the ball to travel the $2.0 \, \text{m}$ to the net after leaving the racket.

(ii) The net is $0.90 \, \text{m}$ high in the centre of the court.

Show by calculation that the ball will go over the net.　　**4**

(9)

23. A toy diving bell consists of an inverted glass bulb, open at one end. The bulb contains a fixed mass of air trapped by water.

The diving bell floats below the surface of the water in a sealed plastic bottle.

The bottle is flexible and can be squeezed.

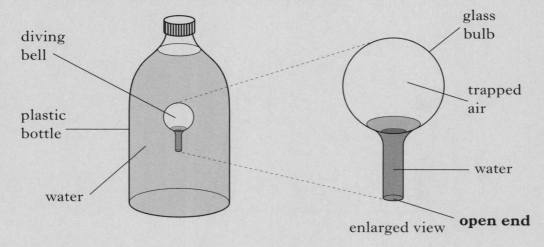

(a) The diving bell has a mass of $2 \cdot 5 \times 10^{-3}$ kg.

Calculate the size of the upthrust (buoyancy force) acting on the bell when it is stationary. **1**

(b) The trapped air inside the diving bell has a volume of $0 \cdot 71$ cm^3, and is at a pressure of $1 \cdot 01 \times 10^5$ Pa. The bottle is now squeezed. This reduces the volume of air trapped inside the bell to $0 \cdot 63$ cm^3. The temperature of the trapped air remains constant.

  (i) Calculate the pressure of the trapped air after the bottle is squeezed.

 (ii) What happens to the volume of **water** inside the bell when the plastic bottle is squeezed?

(iii) Explain why the diving bell sinks when the plastic bottle is squeezed. **5**

(6)

24. An a.c. supply is connected to an oscilloscope.

The trace on the oscilloscope and the oscilloscope settings are shown below.

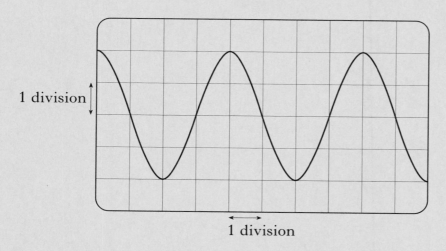
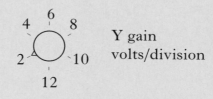
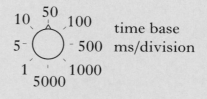

(a) Calculate the r.m.s. voltage of the a.c. supply. **2**

(b) Calculate the frequency of the voltage from the a.c. supply. **2**

(c) A 15 µF capacitor is connected to a 4·5 V battery, a 200 kΩ resistor and a switch as shown in the diagram below. The battery has negligible internal resistance.

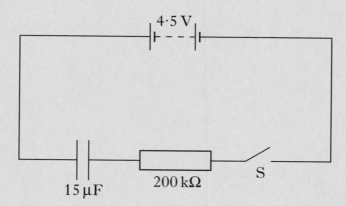

The capacitor is initially uncharged. Switch S is now closed.

(i) What is the voltage across the **resistor** immediately after switch S is closed?

(ii) Sketch a graph showing how the current in the circuit varies with time after switch S is closed. Numerical values on the axes are not required. **2**

(d) (i) Explain why work must be done to charge the capacitor.

(ii) Calculate the energy stored in the capacitor when it is fully charged. **3**

(9)

25. During an experiment to measure the e.m.f. and internal resistance of a cell, the following graph is obtained.

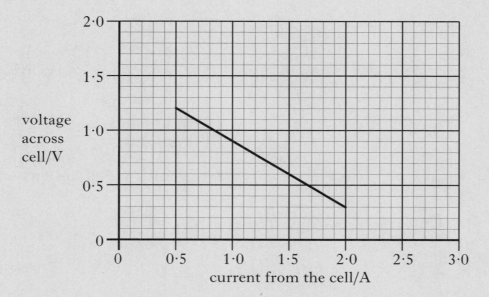

(a) Draw a circuit which could be used to obtain the data for this graph.

(b) (i) What is the value of the e.m.f. of the cell?

(ii) Calculate the internal resistance of the cell.

(c) A length of electrical cable for use at high voltages consists of a steel alloy conductor surrounded by six aluminium conductors as shown in the diagram.

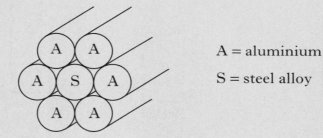

A = aluminium

S = steel alloy

The resistance of one kilometre of each type of conductor is shown below.

Resistance of 1 km of aluminium conductor = $0.60\,\Omega$

Resistance of 1 km of steel alloy conductor = $4.0\,\Omega$

Calculate the resistance of one kilometre of this cable. Give your answer to **three** decimal places.

Marks

25. (continued)

(d) Another type of electrical cable has a much thicker conductor, of the same steel alloy, surrounded by nine aluminium conductors as shown.

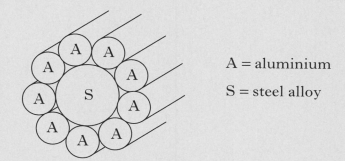

A = aluminium
S = steel alloy

The aluminium conductors have the same cross-section area as those in part (c).

In what way does the resistance of 1 km of this cable differ from the resistance of 1 km of the cable in part (c)? You must justify your answer.

2

(9)

26. The circuit below is used to switch on the motor of a fan in a greenhouse when the temperature in the greenhouse is greater than a certain value.

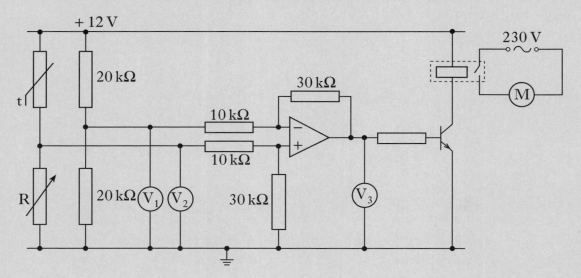

(a) State the mode in which the op-amp is working. **1**

(b) (i) What is the reading on voltmeter V_1?

(ii) At 24 °C, the reading on voltmeter $V_1 = 6·0$ V and the reading on voltmeter $V_2 = 6·2$ V

Calculate the voltage shown on voltmeter V_3. **3**

(c) The resistance of the variable resistor R is now increased.

Explain whether motor M will switch on at a higher or lower temperature than before. **2**

(6)

27. (a) A laser has a mirror at each end of the laser tube as shown in the diagram below.

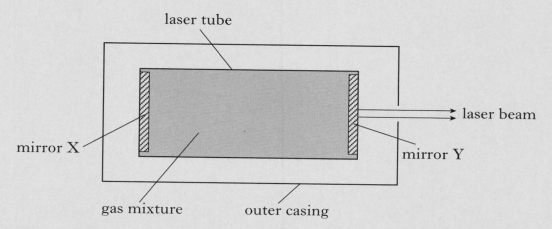

(i) Explain how the mirrors help to amplify the laser light.

(ii) In what way does mirror Y differ from mirror X? **2**

(b) The rating plate on a laser is shown below.

Power of laser beam: 1 mW	Wavelength: 633 nm
230 V a.c.	50 Hz

Calculate the energy of a photon in the beam from this laser. **3**

(c) A photodiode is connected to a voltmeter as shown in the following diagram.

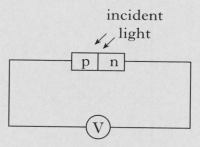

(i) In what mode is the photodiode operating?

(ii) Explain why there is a voltage across the p-n junction when light falls on the photodiode.

(iii) The intensity of light incident on the photodiode increases. Explain why the voltage across the photodiode increases. **3**

(8)

28. The diagram illustrates one method of demonstrating interference of light.

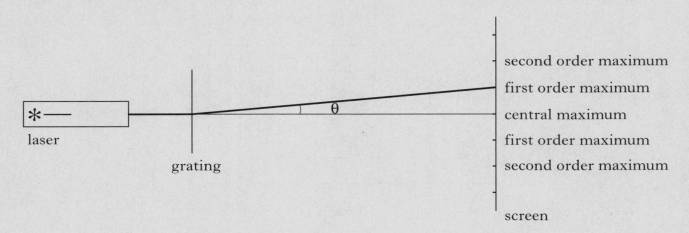

(a) The wavelength of light from the laser is 633 nm, and the separation of lines on the grating is 2.0×10^{-6} m. Calculate the angle θ between the central maximum and first order maximum. **2**

(b) The screen is now moved away from the grating. What effect does this have on the value of θ? **1**

(c) The critical angle for light in a type of red plastic is 41°. Show that the refractive index of this red plastic is 1·52. **1**

28. (continued)

(d) This red plastic is used to make the rear reflector on a bicycle.
The diagrams below show the rear reflector on the bicycle and an enlarged view of the reflector.

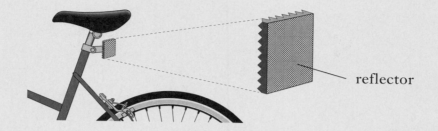

A ray of light from a car headlamp strikes the reflector as shown below.
The refractive index of the red plastic is 1·52.

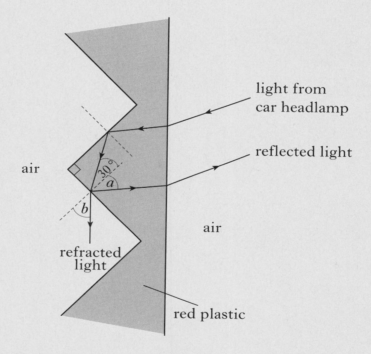

Calculate the values of angles *a* and *b*.

3

(7)

29. Ernest Rutherford arranged for alpha particles to be fired at a thin gold foil. Apparatus similar to that shown below was used.

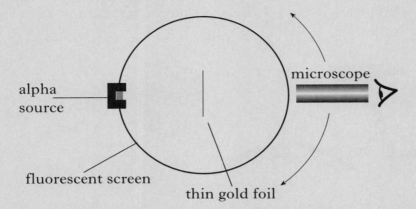

(a) Most of the alpha particles passed straight through the foil.
What does this suggest about the structure of atoms of gold?

(b) A small but significant number of alpha particles were scattered through angles greater than 90°.
What two features about the nucleus does this suggest?

(c) Part of a radioactive decay series is shown below.

$$^{222}_{86}Rn \longrightarrow {}^{218}_{84}Po \longrightarrow {}^{214}_{82}Pb \longrightarrow {}^{214}_{83}Bi \longrightarrow {}^{214}_{84}Po \longrightarrow {}^{210}_{82}Pb$$

(i) Write down one decay from the above series which involves the emission of a beta particle.

(ii) Why is it not possible to tell from the decay series above whether gamma radiation is emitted at any stage?

29. (continued)

(d) A particular nuclear reaction can be described by the following statement.

$$^{3}_{1}H + ^{2}_{1}H \longrightarrow ^{4}_{2}He + ^{1}_{0}n$$

The masses of the nuclides and particle involved in the nuclear reaction are as shown in the table.

	Mass
$^{2}_{1}H$	$3 \cdot 342 \times 10^{-27}$ kg
$^{3}_{1}H$	$5 \cdot 005 \times 10^{-27}$ kg
$^{4}_{2}He$	$6 \cdot 642 \times 10^{-27}$ kg
$^{1}_{0}n$	$1 \cdot 675 \times 10^{-27}$ kg

Calculate the energy available from the above reaction.

3

(8)

[END OF QUESTION PAPER]

2000 HIGHER

X069/301

NATIONAL
QUALIFICATIONS
2000

WEDNESDAY, 31 MAY
9.00 AM – 11.30 AM

**PHYSICS
HIGHER**

Read Carefully

1. All questions should be attempted.

Section A (questions 1 to 20)

2. Check that the answer sheet is for Physics Higher (Section A).
3. Answer the questions numbered 1 to 20 on the answer sheet provided.
4. Fill in the details required on the answer sheet.
5. Rough working, if required, should be done only on this question paper, or on the first two pages of the answer book provided—**not** on the answer sheet.
6. For each of the questions 1 to 20 there is only **one** correct answer and each is worth 1 mark.
7. Instructions as to how to record your answers to questions 1–20 are given on page three.

Section B (questions 21 to 29)

8. Answer questions numbered 21 to 29 in the answer book provided.
9. Fill in the details on the front of the answer book.
10. Enter the question number clearly in the margin of the answer book beside each of your answers to questions 21 to 29.
11. Care should be taken to give an appropriate number of significant figures in the final answers to calculations.

DATA SHEET
COMMON PHYSICAL QUANTITIES

Quantity	Symbol	Value	Quantity	Symbol	Value
Speed of light in vacuum	c	3.00×10^8 m s^{-1}	Mass of electron	m_e	9.11×10^{-31} kg
Magnitude of the charge on an electron	e	1.60×10^{-19} C	Mass of neutron	m_n	1.675×10^{-27} kg
Gravitational acceleration	g	9.8 m s^{-2}	Mass of proton	m_p	1.673×10^{-27} kg
Planck's constant	h	6.63×10^{-34} J s			

REFRACTIVE INDICES

The refractive indices refer to sodium light of wavelength 589 nm and to substances at a temperature of 273 K.

Substance	Refractive index	Substance	Refractive index
Diamond	2·42	Water	1·33
Crown glass	1·50	Air	1·00

SPECTRAL LINES

Element	Wavelength/nm	Colour	Element	Wavelength/nm	Colour
Hydrogen	656	Red	Cadmium	644	Red
	486	Blue-green		509	Green
	434	Blue-violet		480	Blue
	410	Violet		Lasers	
	397	Ultraviolet	Element	Wavelength/nm	Colour
	389	Ultraviolet	Carbon dioxide	9550 / 10590	Infrared
Sodium	589	Yellow	Helium-neon	633	Red

PROPERTIES OF SELECTED MATERIALS

Substance	Density/ kg m^{-3}	Melting Point/ K	Boiling Point/ K
Aluminium	2.70×10^3	933	2623
Copper	8.96×10^3	1357	2853
Ice	9.20×10^2	273	....
Sea Water	1.02×10^3	264	377
Water	1.00×10^3	273	373
Air	1·29	....	....
Hydrogen	9.0×10^{-2}	14	20

The gas densities refer to a temperature of 273 K and a pressure of 1.01×10^5 Pa.

SECTION A

For questions 1 to 20 in this section of the paper, an answer is recorded on the answer sheet by indicating the choice A, B, C, D or E by a stroke made in ink in the appropriate box of the answer sheet—see the example below.

EXAMPLE

The energy unit measured by the electricity meter in your home is the

    A ampere

    B kilowatt-hour

    C watt

    D coulomb

    E volt.

The correct answer to the question is B—kilowatt-hour. Record your answer by drawing a heavy vertical line joining the two dots in the appropriate box on your answer sheet in the column of boxes headed B. The entry on your answer sheet would now look like this:

If after you have recorded your answer you decide that you have made an error and wish to make a change, you should cancel the original answer and put a vertical stroke in the box you now consider to be correct. Thus, if you want to change an answer D to an answer B, your answer sheet would look like this:

If you want to change back to an answer which has already been scored out, you should enter a tick (✓) to the RIGHT of the box of your choice, thus:

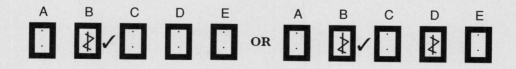

SECTION A

Answer questions 1–20 on the answer sheet.

1. Which of the following is a scalar quantity?
   - A Velocity
   - B Acceleration
   - C Mass
   - D Force
   - E Momentum

2. A woman walks 12 km due North. She then turns round immediately and walks 4 km due South. The total journey takes 4 hours.

   Which row in the following table gives the correct values for her average velocity and average speed?

	Average velocity	Average speed
A	4 km h^{-1} due N	4 km h^{-1}
B	4 km h^{-1} due N	2 km h^{-1}
C	3 km h^{-1} due N	4 km h^{-1}
D	2 km h^{-1} due N	4 km h^{-1}
E	2 km h^{-1} due N	3 km h^{-1}

3. The following velocity-time graph describes the motion of a ball, dropped from rest and bouncing several times.

   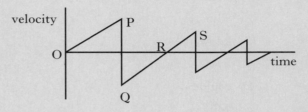

   Which of the following statements is/are true?
   - I The ball hits the ground at P.
   - II The ball is moving upwards between Q and R.
   - III The ball is moving upwards between R and S.

   - A I only
   - B II only
   - C III only
   - D I and II only
   - E I and III only

4. The momentum of a rock of mass 4 kg is 12 kg m s^{-1}.

   The kinetic energy of the rock is
   - A 6 J
   - B 18 J
   - C 36 J
   - D 144 J
   - E 288 J.

5. Density is measured in
   - A N m^{-2}
   - B N m^{-3}
   - C kg m^{3}
   - D kg m^{-2}
   - E kg m^{-3}.

6. The pressure of a fixed mass of gas is 100 kPa at a temperature of −52 °C. The volume of the gas remains constant.

At what temperature would the pressure of the gas be 200 kPa?

A   −26 °C
B   +52 °C
C   +147 °C
D   +169 °C
E   +442 °C

7. The end of a bicycle pump is sealed with a stopper so that the air in the chamber is trapped.

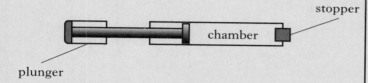

The plunger is now pushed in slowly causing the air in the chamber to be compressed. As a result of this the pressure of the trapped air increases.

Assuming that the temperature remains constant, which of the following explain/s why the pressure increases?

I   The air molecules increase their average speed.

II The air molecules are colliding more often with the walls of the chamber.

III Each air molecule is striking the walls of the chamber with greater force.

A   II only
B   III only
C   I and II only
D   I and III only
E   I, II and III

8. One volt is

A   one coulomb per joule
B   one joule coulomb
C   one joule per coulomb
D   one joule per second
E   one coulomb per second.

9. In the following circuit the reading on the voltmeter is zero.

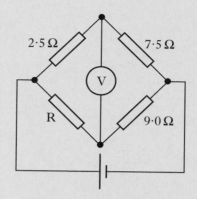

The resistance of resistor R is

A   0·33 Ω
B   0·48 Ω
C   2·1 Ω
D   3·0 Ω
E   27 Ω.

[Turn over

10. The circuits below have identical a.c. supplies which are set at a frequency of 200 Hz. A current is registered on each of the ammeters A_1 and A_2.

constant amplitude
variable frequency

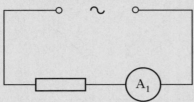

constant amplitude
variable frequency

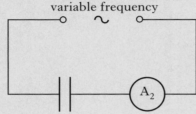

The frequency of each a.c. supply is now increased to 500 Hz.

What happens to the readings on ammeters A_1 and A_2?

	A_1	A_2
A	increases	decreases
B	decreases	increases
C	no change	no change
D	no change	decreases
E	no change	increases

11. A student makes the following statements about ideal op-amps.

  I An op-amp used in the inverting mode inverts the input signal.

  II The gain equation for the inverting mode is
$$\frac{V_o}{V_1} = -\frac{R_1}{R_f}$$
  where the symbols have their usual meanings.

  III An op-amp used in the differential mode amplifies the sum of its two input voltages.

Which of the above statements is/are correct?

A I only

B II only

C III only

D I and II only

E I, II and III

12. An op-amp circuit is connected as shown below.

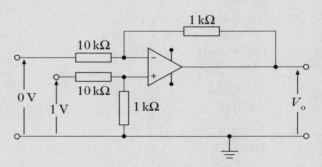

What is the value of the output voltage V_o?

A   10 V

B   0·1 V

C   0 V

D   −0·1 V

E   −10 V

13. The circuit below is used to generate square waves. The amplitude of the alternating input voltage is 6 V.

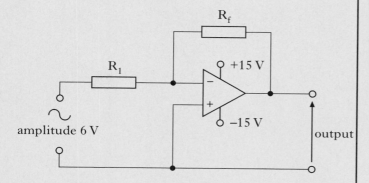

Which values for resistors R_1 and R_f will produce an approximate square wave output?

	R_1	R_f
A	1 kΩ	10 kΩ
B	5 kΩ	10 kΩ
C	10 kΩ	10 kΩ
D	10 kΩ	5 kΩ
E	10 kΩ	1 kΩ

14. Waves from coherent sources, S_1 and S_2, produce an interference pattern. Maxima of intensity are detected at the positions shown below.

The path difference $S_1K - S_2K$ is 154 mm. The wavelength of the waves is

A 15·4 mm

B 25·7 mm

C 28·0 mm

D 30·8 mm

E 34·2 mm.

15. When white light passes through a grating, maxima of intensity are produced on a screen, as shown below. The central maximum is white. Continuous spectra are obtained at positions P and Q.

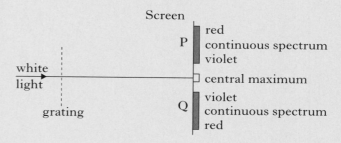

In the continuous spectra, violet is observed closest to the central maximum.
Which of the following statements is/are true?

I Violet light has the shortest wavelength of all the visible radiations.

II Violet light has the longest wavelength of all the visible radiations.

III Violet light travels faster through air than the other visible radiations.

A I only

B II only

C III only

D I and III only

E II and III only

16. A ray of light passes from air into a substance that has a refractive index of 2·0. In air, the light has a wavelength λ and frequency f.

Which row in the following table gives the wavelength and frequency of the light in the substance?

	Wavelength	Frequency
A	λ	f
B	$\lambda/2$	$f/2$
C	$\lambda/2$	f
D	2λ	$2f$
E	2λ	f

[Turn over

17. The diagram below shows a ray of red light passing through a semicircular block of glass.

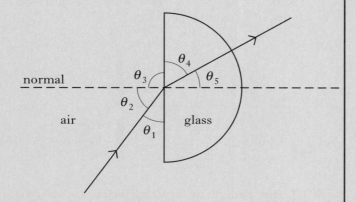

The refractive index of the glass for this light can be calculated from

A $\dfrac{\sin \theta_3}{\sin \theta_4}$

B $\dfrac{\sin \theta_1}{\sin \theta_4}$

C $\dfrac{\sin \theta_2}{\sin \theta_5}$

D $\dfrac{\sin \theta_2}{\sin \theta_4}$

E $\dfrac{\sin \theta_1}{\sin \theta_5}$.

18. The statement below represents a nuclear reaction.

$$^{235}_{92}U + ^{1}_{0}n \rightarrow ^{92}_{36}Kr + ^{141}_{56}Ba + ^{1}_{0}n + ^{1}_{0}n + ^{1}_{0}n$$

This is an example of

A nuclear fusion

B alpha particle emission

C beta particle emission

D spontaneous nuclear fission

E induced nuclear fission.

19. A radioactive source that emits gamma radiation is kept in a large container. A count rate of 160 counts per minute, after correction for background radiation, is recorded outside the container.

The container is to be shielded so that the corrected count rate at the same point outside the container is no more than 10 counts per minute.

Lead and water are available as shielding materials. For this source, the half-value thickness of lead is 11 mm and the half-value thickness of water is 110 mm.

Which of the following shielding arrangements will comply with the above requirement?

A 40 mm of lead only

B 33 mm of lead plus 110 mm of water

C 20 mm of lead plus 220 mm of water

D 11 mm of lead plus 275 mm of water

E 10 mm of lead plus 330 mm of water

20. The diagram below represents possible energy levels of an atom.

P ——————————— $-5 \cdot 2 \times 10^{-19}$ J

Q ——————————— $-9 \cdot 0 \times 10^{-19}$ J

R ——————————— $-16 \cdot 4 \times 10^{-19}$ J

S ——————————— $-24 \cdot 6 \times 10^{-19}$ J

Which of the following statements is/are true?

I There are four emission lines in the spectrum produced as a result of transitions between the energy levels shown.

II The radiation emitted with the shortest wavelength is produced by an electron falling from level P to level S.

III The zero energy level in an energy level diagram is known as the ionisation level.

A I and II only

B I and III only

C II and III only

D III only

E I, II and III

SECTION B

Write your answers to questions 21 to 29 in the answer book.

21. At a funfair, a prize is awarded if a coin is tossed into a small dish. The dish is mounted on a shelf above the ground as shown.

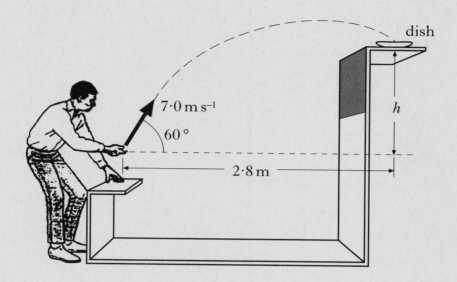

A contestant projects the coin with a speed of 7·0 m s⁻¹ at an angle of 60° to the horizontal. When the coin leaves his hand, the **horizontal distance** between the coin and the dish is 2·8 m. The coin lands in the dish.

The effect of air friction on the coin may be neglected.

(a) Calculate:

   (i) the horizontal component of the initial velocity of the coin;

   (ii) the vertical component of the initial velocity of the coin. **2**

(b) Show that the time taken for the coin to reach the dish is 0·8 s. **1**

(c) What is the height, h, of the shelf above the point where the coin leaves the contestant's hand? **2**

(d) How does the value of the kinetic energy of the coin when it enters the dish compare with the kinetic energy of the coin just as it leaves the contestant's hand?

Justify your answer. **2**

(7)

[Turn over

22. The apparatus shown below is used to test concrete pipes.

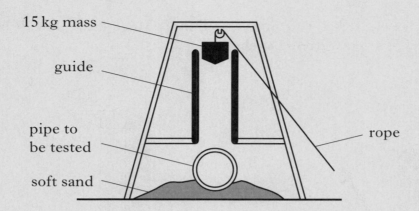

When the rope is released, the 15 kg mass is dropped and falls freely through a distance of 2·0 m on to the pipe.

(a) In one test, the mass is dropped on to an uncovered pipe.

   (i) Calculate the speed of the mass just before it hits the pipe.

   (ii) When the 15 kg mass hits the pipe the mass is brought to rest in a time of 0·02 s. Calculate the size and direction of the average unbalanced force on the **pipe**. 5

(b) The same 15 kg mass is now dropped through the same distance on to an identical pipe which is covered with a thick layer of soft material.

   Describe and explain the effect this layer has on the size of the average unbalanced force on the pipe. 2

(c) Two 15 kg masses, X and Y, shaped as shown, are dropped through the same distance on to identical uncovered concrete pipes.

When the masses hit the pipes, the masses are brought to rest in the same time.

Which mass causes more damage to a pipe?

Explain your answer in terms of pressure. 2

(9)

23. A sonar detector is attached to the bottom of a fresh water loch by a vertical cable as shown.

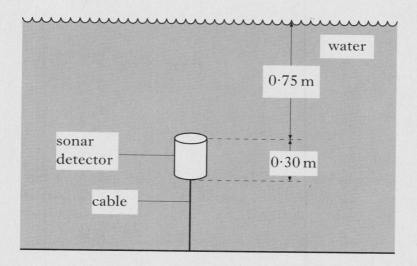

The detector has a mass of 100 kg. Each end of the detector has an area of 0·40 m². Atmospheric pressure is 101 000 Pa.

(a) The total pressure on the top of the detector is 108 350 Pa.

Show that the total pressure on the bottom of the detector is 111 290 Pa. **2**

(b) Calculate the upthrust on the detector. **3**

(c) The sonar detector is now attached, as before, to the bottom of a **sea water** loch. The top of the detector is again 0·75 m below the surface of the water.

How does the size of the upthrust on the detector now compare with your answer to (b)?

You must justify your answer. **2**

(7)

[Turn over

24. (a) In an experiment to measure the capacitance of a capacitor, a student sets up the following circuit.

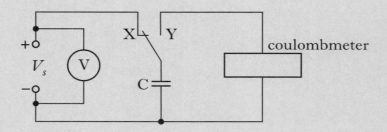

When the switch is in position X, the capacitor charges up to the supply voltage, V_s. When the switch is in position Y, the coulombmeter indicates the charge stored by the capacitor.

The student records the following measurements and uncertainties.

Reading on voltmeter = $(2\cdot56 \pm 0\cdot01)$ V
Reading on coulombmeter = $(32 \pm 1)\,\mu$C

Calculate the value of the capacitance and the percentage uncertainty in this value. You must give the answer in the form

value ± percentage uncertainty.

(b) The student designs the circuit shown below to switch off a lamp after a certain time.

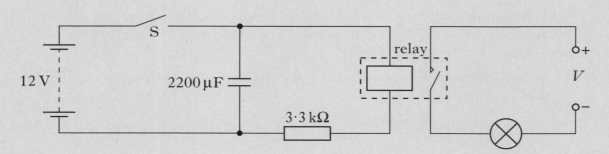

The 12 V battery has negligible internal resistance.

The relay contacts are normally open. When there is a current in the relay coil the contacts close and complete the lamp circuit.

Switch S is initially closed and the lamp is on.

(i) What is the maximum energy stored in the capacitor?

(ii) (A) Switch S is now opened. Explain why the lamp stays lit for a few seconds.

(B) The 2200 μF capacitor is replaced with a 1000 μF capacitor.

Describe and explain the effect of this change on the operation of the circuit.

25. A photodiode is connected in a circuit as shown below.

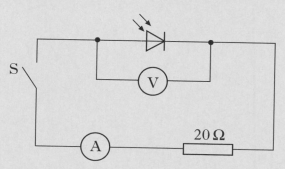

Switch S is open.

Light is shone on to the photodiode.

A reading is obtained on the voltmeter.

(a) (i) State the mode in which the photodiode is operating.

    (ii) Describe the effect of light on the material of which the photodiode is made.

    (iii) The intensity of the light on the photodiode is increased.

    What happens to the reading on the voltmeter? **3**

(b) Light of a constant intensity is shone on to the photodiode in the circuit shown above.

The following measurements are obtained with S open and then with S closed.

	S open	S closed
reading on voltmeter/V	0·508	0·040
reading on ammeter/mA	0·00	2·00

    (i) What is the value of the e.m.f. produced by the photodiode for this light intensity?

    (ii) Calculate the internal resistance of the photodiode for this light intensity. **3**

(c) In the circuit above, the 20 Ω resistor is now replaced with a 10 Ω resistor.

The intensity of the light is unchanged.

The following measurements are obtained.

	S open	S closed
reading on voltmeter/V	0·508	0·021

Explain why the reading on the voltmeter, when S is closed, is smaller than the corresponding reading in part (b). **2**

(8)

26. A circuit is set up as shown below. The amplitude of the output voltage of the a.c. supply is kept constant.

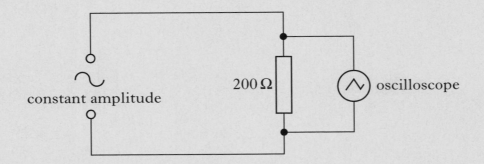

The settings of the controls on the oscilloscope are as follows:

y-gain setting = 5 V/division
time-base setting = 2·5 ms/division

The following trace is displayed on the oscilloscope screen.

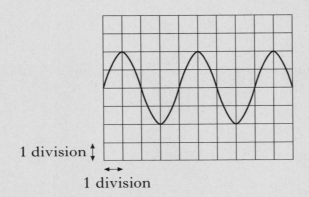

(a) (i) Calculate the frequency of the output from the a.c. supply.
(ii) Calculate the **r.m.s. current** in the 200 Ω resistor.

5

26. (continued)

(b) A diode is now connected in the circuit as shown below.

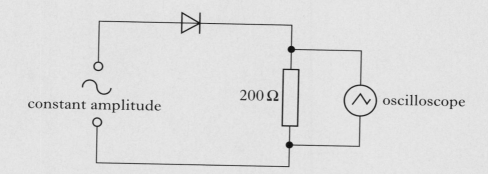

The settings on the controls of the oscilloscope remain unchanged.

Connecting the diode in the circuit causes **changes** to the original trace displayed on the oscilloscope screen. The new trace is shown below.

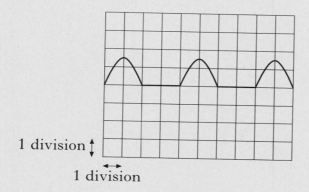

1 division ↕
1 division ↔

Describe and explain the changes to the original trace.

2

(7)

[Turn over

Marks

27. A student is investigating the effect that a semicircular glass block has on a ray of monochromatic light.

She observes that at point X the incident ray splits into two rays:

T — a transmitted ray
R — a reflected ray.

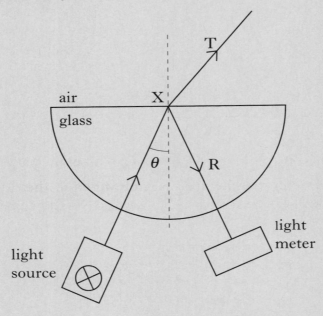

The student uses a light meter to measure the intensity of ray R as angle θ is changed.

(a) State what is meant by the *intensity* of a radiation. **1**

(b) Explain why, as angle θ is changed, it is important to keep the light meter at a constant distance from point X for each measurement of intensity. **1**

27. (continued)

(c) The graph below is obtained from the student's results.

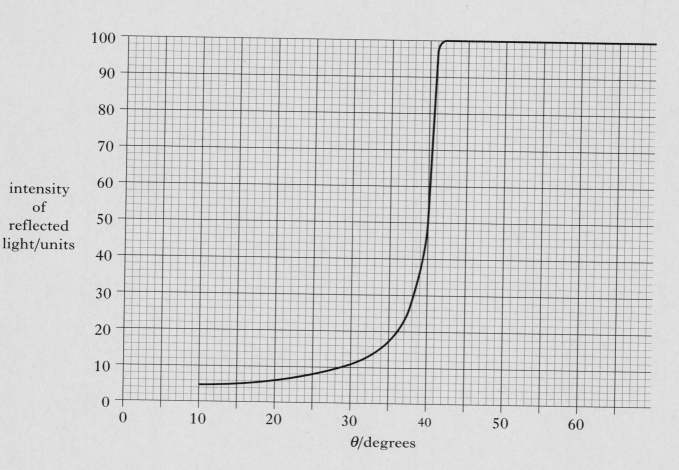

(i) What is the value of the critical angle in the glass for this light?
(ii) Calculate the refractive index of the glass for this light.
(iii) As the angle θ is increased, what happens to the intensity of ray T?

4

(6)

[Turn over

Marks

28. (*a*) The apparatus shown below is used to investigate photoelectric emission from a metal surface when electromagnetic radiation is shone on the surface.

The intensity and frequency of the incident radiation can be varied as required.

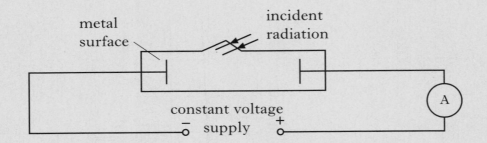

(i) Explain what is meant by *photoelectric emission* from a metal.

(ii) What is the name given to the minimum frequency of the radiation that produces a current in the circuit?

(iii) A particular source of radiation produces a current in the circuit.

Explain why the current in the circuit increases as the intensity of the incident radiation increases.

4

(*b*) A semiconductor chip is used to store information. The information can only be erased by exposing the chip to ultraviolet radiation for a period of time.

The following data is provided.

Frequency of ultraviolet radiation used	$= 9 \cdot 0 \times 10^{14}\,\text{Hz}$
Minimum intensity of ultraviolet radiation required at the chip	$= 25\,\text{W m}^{-2}$
Area of the chip exposed to radiation	$= 1 \cdot 8 \times 10^{-9}\,\text{m}^2$
Time taken to erase the information	$= 15\,\text{minutes}$
Energy of radiation needed to erase the information	$= 40 \cdot 5\,\mu\text{J}$

(i) Calculate the energy of a photon of the ultraviolet radiation used.

(ii) Calculate the number of photons of the ultraviolet radiation required to erase the information.

(iii) Sunlight of intensity $25\,\text{W m}^{-2}$, at the chip, can also be used to erase the information.

State whether the time taken to erase the information is greater than, equal to or less than 15 minutes.

You must justify your answer.

5

(9)

29. Radium (Ra) decays to radon (Rn) by the emission of an alpha particle. Some energy is also released by this decay.

The decay is represented by the statement shown below.

$$^{226}_{88}Ra \longrightarrow {}^{x}_{y}Rn + {}^{4}_{2}He$$

The masses of the nuclides involved are as follows.

Mass of $^{226}_{88}Ra$ = 3.75428×10^{-25} kg

Mass of $^{x}_{y}Rn$ = 3.68771×10^{-25} kg

Mass of $^{4}_{2}He$ = 6.64832×10^{-27} kg

(a) (i) What are the values of x and y for the nuclide $^{x}_{y}Rn$?

 (ii) Why is energy released by this decay?

 (iii) Calculate the energy released by one decay of this type. **5**

(b) The alpha particle leaves the radium nucleus with a speed of $1.5 \times 10^7 \text{ m s}^{-1}$.

The alpha particle is now accelerated through a potential difference of 25 kV.

Calculate the **final** kinetic energy, in joules, of the alpha particle. **3**

  (8)

[END OF QUESTION PAPER]

2001 HIGHER

X069/301

NATIONAL
QUALIFICATIONS
2001

MONDAY, 4 JUNE
9.00 AM – 11.30 AM

PHYSICS
HIGHER

Read Carefully

1. All questions should be attempted.

Section A (questions 1 to 20)

2. Check that the answer sheet is for Physics Higher (Section A).
3. Answer the questions numbered 1 to 20 on the answer sheet provided.
4. Fill in the details required on the answer sheet.
5. Rough working, if required, should be done only on this question paper, or on the first two pages of the answer book provided—**not** on the answer sheet.
6. For each of the questions 1 to 20 there is only **one** correct answer and each is worth 1 mark.
7. Instructions as to how to record your answers to questions 1–20 are given on page three.

Section B (questions 21 to 29)

8. Answer questions numbered 21 to 29 in the answer book provided.
9. Fill in the details on the front of the answer book.
10. Enter the question number clearly in the margin of the answer book beside each of your answers to questions 21 to 29.
11. Care should be taken to give an appropriate number of significant figures in the final answers to calculations.

DATA SHEET
COMMON PHYSICAL QUANTITIES

Quantity	Symbol	Value	Quantity	Symbol	Value
Speed of light in vacuum	c	3.00×10^8 m s^{-1}	Mass of electron	m_e	9.11×10^{-31} kg
Magnitude of the charge on an electron	e	1.60×10^{-19} C	Mass of neutron	m_n	1.675×10^{-27} kg
Gravitational acceleration	g	9.8 m s^{-2}	Mass of proton	m_p	1.673×10^{-27} kg
Planck's constant	h	6.63×10^{-34} J s			

REFRACTIVE INDICES
The refractive indices refer to sodium light of wavelength 589 nm and to substances at a temperature of 273 K.

Substance	Refractive index	Substance	Refractive index
Diamond	2.42	Water	1.33
Crown glass	1.50	Air	1.00

SPECTRAL LINES

Element	Wavelength/nm	Colour	Element	Wavelength/nm	Colour
Hydrogen	656	Red	Cadmium	644	Red
	486	Blue-green		509	Green
	434	Blue-violet		480	Blue
	410	Violet		Lasers	
	397	Ultraviolet	Element	Wavelength/nm	Colour
	389	Ultraviolet	Carbon dioxide	9550 } 10590 }	Infrared
Sodium	589	Yellow	Helium-neon	633	Red

PROPERTIES OF SELECTED MATERIALS

Substance	Density/ kg m^{-3}	Melting Point/ K	Boiling Point/ K
Aluminium	2.70×10^3	933	2623
Copper	8.96×10^3	1357	2853
Ice	9.20×10^2	273	....
Sea Water	1.02×10^3	264	377
Water	1.00×10^3	273	373
Air	1.29	....	....
Hydrogen	9.0×10^{-2}	14	20

The gas densities refer to a temperature of 273 K and a pressure of 1.01×10^5 Pa.

SECTION A

For questions 1 to 20 in this section of the paper, an answer is recorded on the answer sheet by indicating the choice A, B, C, D or E by a stroke made in ink in the appropriate box of the answer sheet—see the example below.

EXAMPLE

The energy unit measured by the electricity meter in your home is the

    A   ampere

    B   kilowatt-hour

    C   watt

    D   coulomb

    E   volt.

The correct answer to the question is B—kilowatt-hour. Record your answer by drawing a heavy vertical line joining the two dots in the appropriate box on your answer sheet in the column of boxes headed B. The entry on your answer sheet would now look like this:

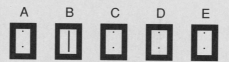

If after you have recorded your answer you decide that you have made an error and wish to make a change, you should cancel the original answer and put a vertical stroke in the box you now consider to be correct. Thus, if you want to change an answer D to an answer B, your answer sheet would look like this:

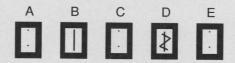

If you want to change back to an answer which has already been scored out, you should enter a tick (✓) to the RIGHT of the box of your choice, thus:

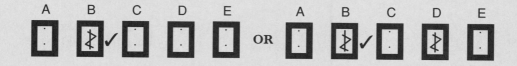

SECTION A

Answer questions 1–20 on the answer sheet.

1. Which one of the following pairs contains one vector quantity and one scalar quantity?

   A Force, kinetic energy
   B Power, speed
   C Displacement, acceleration
   D Work, potential energy
   E Momentum, velocity

2. The diagram below shows the resultant of two vectors.

   Which of the diagrams below shows the vectors which could produce the above resultant?

   A

   B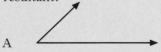

   C

   D

   E

3. A helicopter is **descending** vertically at a constant speed of $3.0\,\text{m s}^{-1}$. A sandbag is released from the helicopter. The sandbag hits the ground $5.0\,\text{s}$ later.

   What was the height of the helicopter above the ground at the time the sandbag was released?

   A $15.0\,\text{m}$
   B $49.0\,\text{m}$
   C $107.5\,\text{m}$
   D $122.5\,\text{m}$
   E $137.5\,\text{m}$

4. A car of mass $900\,\text{kg}$ pulls a caravan of mass $400\,\text{kg}$ along a straight, horizontal road with an acceleration of $2.0\,\text{m s}^{-2}$.

   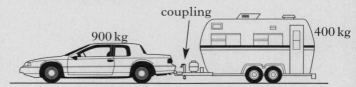

   Assuming that the frictional forces on the caravan are negligible, the tension in the coupling between the car and the caravan is

   A $400\,\text{N}$
   B $500\,\text{N}$
   C $800\,\text{N}$
   D $1800\,\text{N}$
   E $2600\,\text{N}$.

5. A rocket of mass $5.0\,\text{kg}$ is travelling horizontally with a speed of $200\,\text{m s}^{-1}$ when it explodes into two parts. One part of mass $3.0\,\text{kg}$ continues in the original direction with a speed of $100\,\text{m s}^{-1}$. The other part also continues in this same direction. Its speed is

   A $150\,\text{m s}^{-1}$
   B $200\,\text{m s}^{-1}$
   C $300\,\text{m s}^{-1}$
   D $350\,\text{m s}^{-1}$
   E $700\,\text{m s}^{-1}$.

6. A block floats in water and two other liquids X and Y at the levels shown.

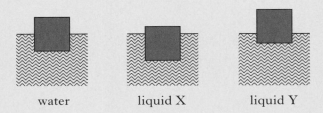

water   liquid X   liquid Y

Which of the following statements is/are correct?

I The density of the material of the block is less than the density of water.

II The density of liquid X is less than the density of water.

III The density of liquid X is greater than the density of liquid Y.

A I only
B II only
C I and II only
D I and III only
E II and III only

7. Ice at −10 °C is heated until it becomes water at 80 °C.

The temperature change on the kelvin scale is

A 70 K
B 90 K
C 343 K
D 363 K
E 636 K.

8. In the diagrams below, each resistor has a resistance of 1·0 ohm.

Select the combination which has the **least** value of effective resistance between the terminals X and Y.

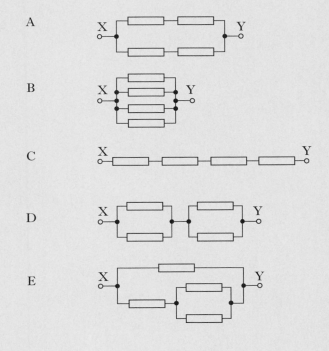

9. In the following circuit, the supply has negligible internal resistance.

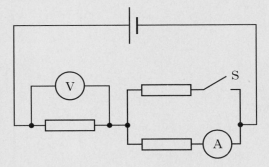

Switch S is now closed.

Which row in the table shows the effect on the ammeter and voltmeter readings?

	Ammeter reading	Voltmeter reading
A	increases	increases
B	increases	decreases
C	decreases	decreases
D	decreases	increases
E	decreases	remains the same

10. A supply with a sinusoidally alternating output of 6·0 V r.m.s. is connected to a 3·0 Ω resistor.

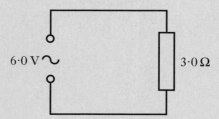

Which row in the following table shows the peak voltage across the resistor and the peak current in the circuit?

	Peak voltage/V	Peak current/A
A	$6\sqrt{2}$	$2\sqrt{2}$
B	$6\sqrt{2}$	2
C	6	2
D	$6\sqrt{2}$	$\dfrac{1}{2\sqrt{2}}$
E	6	$2\sqrt{2}$

11. A resistor and an ammeter are connected to a signal generator having an output of constant amplitude and variable frequency.

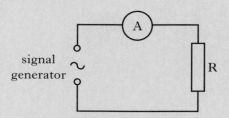

Which of the following graphs shows the correct relationship between the current I in the resistor and the output frequency f of the signal generator?

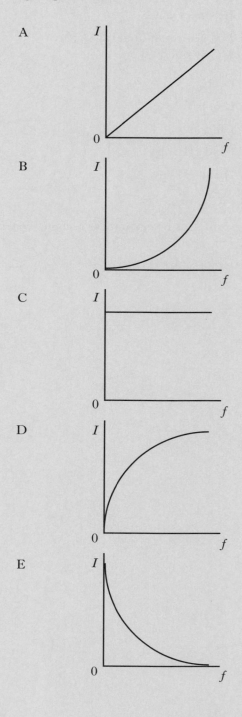

12. Which of the following statements is/are true for an ideal op-amp?

   I It has infinite input resistance.

   II Both input pins are at the same potential.

   III The input current to the op-amp is zero.

   A   I only
   B   II only
   C   I and II only
   D   II and III only
   E   I, II and III

13. An op-amp circuit is shown in the diagram.

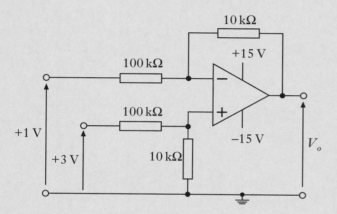

What is the output voltage V_o?

   A   −20 V
   B   −2 V
   C   −0·2 V
   D   0·2 V
   E   20 V

14. The energy of a water wave depends on its

   A   speed
   B   wavelength
   C   frequency
   D   period
   E   amplitude.

15. S_1 and S_2 are sources of coherent waves which produce an interference pattern along the line XY.

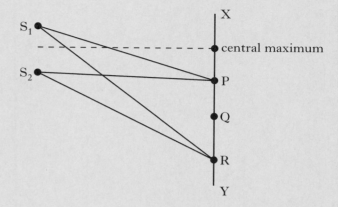

The first maximum occurs at P, where $S_1P = 20$ cm and $S_2P = 18$ cm.

For the third maximum, at R, the path difference $(S_1R - S_2R)$ is

   A   3 cm
   B   4 cm
   C   5 cm
   D   6 cm
   E   8 cm.

16. The spectrum of white light from a filament lamp may be viewed using a prism or a grating.

A student, asked to compare the spectra formed by the two methods, made the following statements.

   I The prism produces a spectrum by refraction. The grating produces a spectrum by interference.

   II The spectrum formed by the prism shows all the wavelengths present in the white light. The spectrum formed by the grating shows only a few specific wavelengths.

   III The prism produces a single spectrum. The grating produces more than one spectrum.

Which of the above statements is/are true?

   A   I only
   B   II only
   C   I and II only
   D   I and III only
   E   I, II and III

17. Red light passes from air into water.

What happens to the speed and frequency of the light when it enters the water?

	Speed	Frequency
A	increases	increases
B	increases	stays constant
C	decreases	stays constant
D	decreases	decreases
E	stays constant	decreases

18. The intensity of light from a point source is $20\,\text{W m}^{-2}$ at a distance of $5 \cdot 0\,\text{m}$ from the source.

What is the intensity of the light at a distance of $25\,\text{m}$ from the source?

A $0 \cdot 032\,\text{W m}^{-2}$

B $0 \cdot 80\,\text{W m}^{-2}$

C $1 \cdot 2\,\text{W m}^{-2}$

D $4 \cdot 0\,\text{W m}^{-2}$

E $100\,\text{W m}^{-2}$

19. Ultraviolet radiation causes the emission of photoelectrons from a zinc plate.

The intensity of the ultraviolet radiation is increased. Which row in the following table shows the effect of this change?

	Maximum kinetic energy of a photoelectron	Number of photoelectrons per second
A	increases	no change
B	no change	increases
C	no change	no change
D	increases	increases
E	decreases	increases

20. Under certain conditions, a nucleus of nitrogen absorbs an alpha particle to form the nucleus of another element and releases a single particle.

Which one of the following statements correctly describes this process?

A $^{14}_{7}\text{N} + ^{3}_{2}\text{He} \rightarrow ^{16}_{9}\text{F} + ^{1}_{0}\text{n}$

B $^{14}_{7}\text{N} + ^{4}_{2}\text{He} \rightarrow ^{17}_{10}\text{N} + ^{0}_{-1}\text{e}$

C $^{14}_{7}\text{N} + ^{3}_{2}\text{He} \rightarrow ^{16}_{8}\text{O} + ^{1}_{1}\text{p}$

D $^{14}_{7}\text{N} + ^{4}_{2}\text{He} \rightarrow ^{18}_{9}\text{F} + 2\,^{0}_{-1}\text{e}$

E $^{14}_{7}\text{N} + ^{4}_{2}\text{He} \rightarrow ^{17}_{8}\text{O} + ^{1}_{1}\text{p}$

SECTION B

Write your answers to questions 21 to 29 in the answer book.

Marks

21. (a) A box of mass 18 kg is at rest on a horizontal frictionless surface.
    A force of 4·0 N is applied to the box at an angle of 26° to the horizontal.

    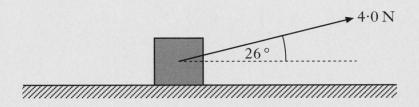

    (i) Show that the horizontal component of this force is 3·6 N.

    (ii) Calculate the acceleration of the box along the horizontal surface.

    (iii) Calculate the horizontal distance travelled by the box in a time of 7·0 s.

    5

    (b) The box is replaced at rest at its starting position.

    The force of 4·0 N is now applied to the box at an angle of less than 26° to the horizontal.

    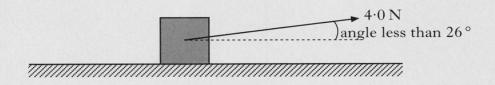

    The force is applied for a time of 7·0 s as before.

    How does the distance travelled by the box compare with your answer to part (a)(iii)?

    You must justify your answer.

    2

    (7)

[Turn over

Marks

22. (a) In an experiment to find the density of air, a student first measures the mass of a flask full of air as shown below.

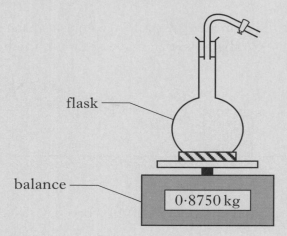

The air is now removed from the flask and the mass of the evacuated flask measured.

This procedure is repeated a number of times and the following table of measurements is obtained.

	Experiment number					
	1	2	3	4	5	6
Mass of flask and air/kg	0·8750	0·8762	0·8748	0·8755	0·8760	0·8757
Mass of evacuated flask/kg	0·8722	0·8736	0·8721	0·8728	0·8738	0·8732
Mass of air removed/kg						

The volume of the flask is measured as $2·0 \times 10^{-3}$ m^3.

(i) Copy and complete the **bottom row** of the table.

(ii) Calculate the mean mass of air removed from the flask **and** the random uncertainty in this mean. Express the mean mass and the random uncertainty in kilograms.

(iii) Use these measurements to calculate the density of air.

(iv) Another student carries out the same experiment using a flask of larger volume.

Explain why this is a better design for the experiment. **6**

22. (continued)

(b) The cylinder of a bicycle pump has a length of 360 mm as shown in the diagram.

The outlet of the pump is sealed.

The piston is pushed inwards until it is 160 mm from the outlet.

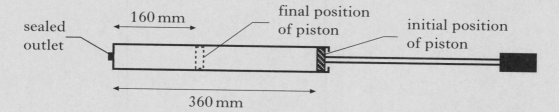

The initial pressure of the air in the pump is 1.0×10^5 Pa.

(i) Assuming that the temperature of the air trapped in the cylinder remains constant, calculate the final pressure of the trapped air.

(ii) State one other assumption you have made for this calculation.

(iii) Use the kinetic model to explain what happens to the pressure of the trapped air as its volume decreases.

5

(11)

[Turn over

Marks

23. Beads of liquid moving at high speed are used to move threads in modern weaving machines.

    (a) In one design of machine, beads of water are accelerated by jets of air as shown in the diagram.

    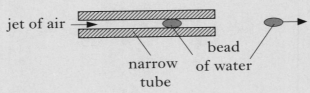

    Each bead has a mass of 2.5×10^{-5} kg.

    When designing the machine, it was estimated that each bead of water would start from rest and experience a constant unbalanced force of 0.5 N for a time of 3.0 ms.

    (i) Calculate:

    (A) the impulse on a bead of water;

    (B) the speed of the bead as it emerges from the tube.

    (ii) In practice the force on a bead varies.

    The following graph shows how the actual unbalanced force exerted on each bead of water varies with time.

    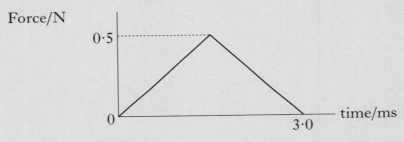

    Use information from this graph to show that the bead leaves the tube with a speed equal to half of the value calculated in part (i)(B). **6**

    (b) Another design of machine uses beads of oil and two metal plates X and Y.

    The potential difference between these plates is 5.0×10^3 V.

    Each bead of oil has a mass of 4.0×10^{-5} kg and is given a negative charge of 6.5×10^{-6} C.

    The bead accelerates from rest at plate X and passes through a hole in plate Y.

    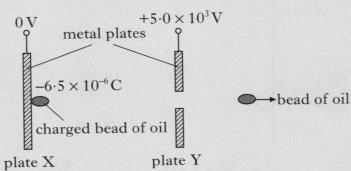

    Neglecting air friction, calculate the speed of the bead at plate Y. **3**

    (9)

24. (a) The following circuit is used to measure the e.m.f. and the internal resistance of a battery.

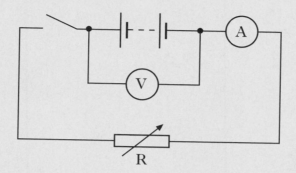

Readings of current and potential difference from this circuit are used to produce the following graph.

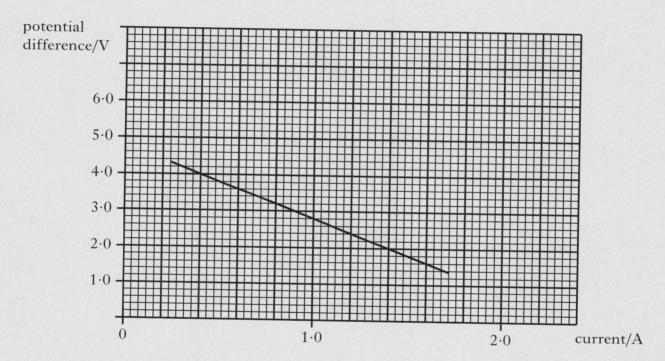

Use information from the graph to find:

(i) the e.m.f. of the battery, in volts;

(ii) the internal resistance of the battery. **3**

(b) A car battery has an e.m.f. of 12 V and an internal resistance of $0.050\,\Omega$.

(i) Calculate the short circuit current for this battery.

(ii) The battery is now connected in series with a lamp. The resistance of the lamp is $2.5\,\Omega$. Calculate the power dissipated in the lamp. **5**

(8)

[Turn over

25. (*a*) The following diagram shows a circuit that is used to investigate the charging of a capacitor.

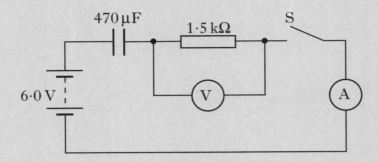

The capacitor is initially uncharged.

The capacitor has a capacitance of 470 μF and the resistor has a resistance of 1·5 kΩ.

The battery has an e.m.f. of 6·0 V and negligible internal resistance.

(i) Switch S is now closed. What is the initial current in the circuit?

(ii) How much energy is stored in the capacitor when it is fully charged?

(iii) What change could be made to this circuit to ensure that the **same** capacitor stores **more** energy? 5

(*b*) A capacitor is used to provide the energy for an electronic flash in a camera.

When the flash is fired, $6·35 \times 10^{-3}$ J of the stored energy is emitted as light.

The mean value of the frequency of photons of light from the flash is $5·80 \times 10^{14}$ Hz.

Calculate the number of photons emitted in each flash of light. 3

(8)

26. (a) An op-amp is connected in a circuit as shown below.

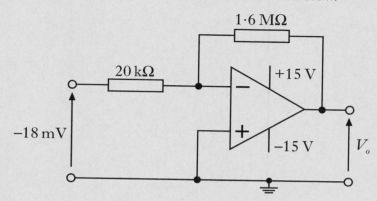

(i) In which mode is the op-amp operating?

(ii) A voltage of −18 mV is connected to the input. Calculate the output voltage V_o.

(iii) The supply voltage is now reduced from ±15 V to ±12 V.

State any effect this change has on the output voltage. You must justify your answer.

(b) A student connects an op-amp as shown in the following diagram. An alternating voltage of peak value 5·0 V is connected to the input as shown.

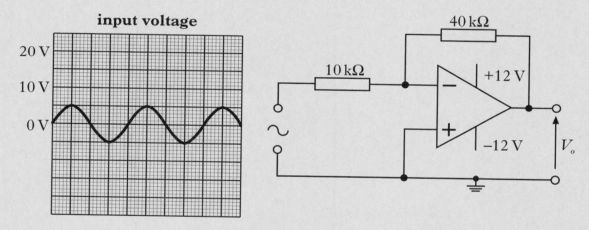

The sketch below shows the student's attempt to draw the corresponding output voltage.

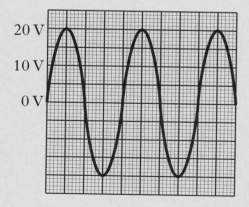

State the **two** mistakes in the student's sketch.

27. (a) Light of wavelength 486×10^{-9} m is viewed using a grating with a slit spacing of $2 \cdot 16 \times 10^{-6}$ m.

Calculate the angle between the central maximum and the second order maximum.

(b) A ray of monochromatic light passes from air into a block of glass as shown.

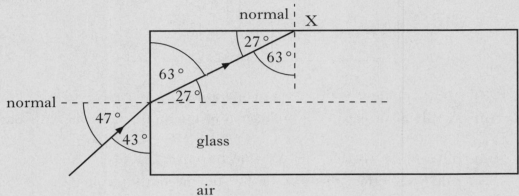

(i) Using information from the diagram, show that the refractive index of the glass for this light is 1·61.

(ii) Show by calculation whether the ray is totally internally reflected at point X.

28. (a) In a laser, the light is produced by stimulated emission of radiation.

Explain the term "stimulated emission" by making reference to the energy levels in atoms.

2

(b) A laser beam is shone on to a screen which is marked with a grid.

The beam produces a uniformly lit spot of radius 5.00×10^{-4} m as shown.

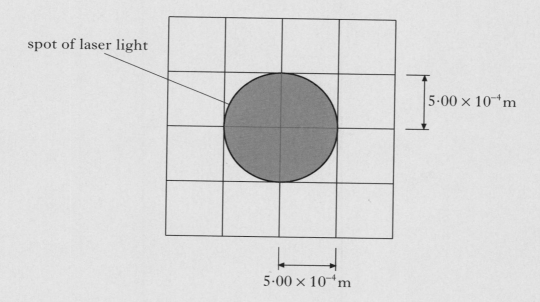

(i) The intensity of the spot of light on the screen is 1020 W m^{-2}.

Calculate the power of the laser beam.

(ii) The distance between the screen and the laser is now doubled.

State how the radius of the spot now compares with the one shown in the diagram.

You must justify your answer.

5

(7)

[Turn over

29. (*a*) The following statement represents a nuclear reaction.

$$^{239}_{94}\text{Pu} + ^{1}_{0}\text{n} \longrightarrow ^{137}_{52}\text{Te} + ^{100}_{42}\text{Mo} + 3\,^{1}_{0}\text{n} + \text{energy}$$

The total mass of the particles before the reaction is $3 \cdot 9842 \times 10^{-25}$ kg and the total mass of the particles after the reaction is $3 \cdot 9825 \times 10^{-25}$ kg.

  (i) State and explain whether this reaction is spontaneous or induced.

  (ii) Calculate the energy, in joules, released by this reaction.

(*b*) A radioactive source is used to irradiate a sample of tissue of mass $0 \cdot 50$ kg.

The tissue absorbs $9 \cdot 6 \times 10^{-5}$ J of energy from the radiation emitted from the source.

The radiation has a quality factor of 1.

  (i) Calculate the absorbed dose received by the tissue.

  (ii) Calculate the dose equivalent received by the tissue.

  (iii) Placing a sheet of lead between the source and the tissue would have reduced the dose received by the tissue.

The half-value thickness of lead for this radiation is 40 mm.

Calculate the thickness of lead which would have limited the absorbed dose to one eighth of the value calculated in part (*b*)(i).

Marks

3

5

(8)

[*END OF QUESTION PAPER*]

2002 HIGHER

X069/301

| NATIONAL QUALIFICATIONS 2002 | WEDNESDAY, 22 MAY 1.00 PM – 3.30 PM | PHYSICS HIGHER |

Read Carefully

1. All questions should be attempted.

Section A (questions 1 to 20)

2. Check that the answer sheet is for Physics Higher (Section A).
3. Answer the questions numbered 1 to 20 on the answer sheet provided.
4. Fill in the details required on the answer sheet.
5. Rough working, if required, should be done only on this question paper, or on the first two pages of the answer book provided—**not** on the answer sheet.
6. For each of the questions 1 to 20 there is only **one** correct answer and each is worth 1 mark.
7. Instructions as to how to record your answers to questions 1–20 are given on page three.

Section B (questions 21 to 30)

8. Answer questions numbered 21 to 30 in the answer book provided.
9. Fill in the details on the front of the answer book.
10. Enter the question number clearly in the margin of the answer book beside each of your answers to questions 21 to 30.
11. Care should be taken to give an appropriate number of significant figures in the final answers to calculations.

DATA SHEET
COMMON PHYSICAL QUANTITIES

Quantity	Symbol	Value	Quantity	Symbol	Value
Speed of light in vacuum	c	3.00×10^8 m s^{-1}	Mass of electron	m_e	9.11×10^{-31} kg
Magnitude of the charge on an electron	e	1.60×10^{-19} C	Mass of neutron	m_n	1.675×10^{-27} kg
Gravitational acceleration	g	9.8 m s^{-2}	Mass of proton	m_p	1.673×10^{-27} kg
Planck's constant	h	6.63×10^{-34} J s			

REFRACTIVE INDICES
The refractive indices refer to sodium light of wavelength 589 nm and to substances at a temperature of 273 K.

Substance	Refractive index	Substance	Refractive index
Diamond	2.42	Water	1.33
Crown glass	1.50	Air	1.00

SPECTRAL LINES

Element	Wavelength/nm	Colour	Element	Wavelength/nm	Colour
Hydrogen	656	Red	Cadmium	644	Red
	486	Blue-green		509	Green
	434	Blue-violet		480	Blue
	410	Violet	Lasers		
	397	Ultraviolet	Element	Wavelength/nm	Colour
	389	Ultraviolet	Carbon dioxide	9550 } 10590 }	Infrared
Sodium	589	Yellow	Helium-neon	633	Red

PROPERTIES OF SELECTED MATERIALS

Substance	Density/ kg m^{-3}	Melting Point/ K	Boiling Point/ K
Aluminium	2.70×10^3	933	2623
Copper	8.96×10^3	1357	2853
Ice	9.20×10^2	273	....
Sea Water	1.02×10^3	264	377
Water	1.00×10^3	273	373
Air	1.29	....	....
Hydrogen	9.0×10^{-2}	14	20

The gas densities refer to a temperature of 273 K and a pressure of 1.01×10^5 Pa.

SECTION A

For questions 1 to 20 in this section of the paper, an answer is recorded on the answer sheet by indicating the choice A, B, C, D or E by a stroke made in ink in the appropriate box of the answer sheet—see the example below.

EXAMPLE

The energy unit measured by the electricity meter in your home is the

  A ampere

  B kilowatt-hour

  C watt

  D coulomb

  E volt.

The correct answer to the question is B—kilowatt-hour. Record your answer by drawing a heavy vertical line joining the two dots in the appropriate box on your answer sheet in the column of boxes headed B. The entry on your answer sheet would now look like this:

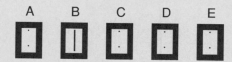

If after you have recorded your answer you decide that you have made an error and wish to make a change, you should cancel the original answer and put a vertical stroke in the box you now consider to be correct. Thus, if you want to change an answer D to an answer B, your answer sheet would look like this:

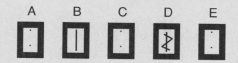

If you want to change back to an answer which has already been scored out, you should enter a tick (✓) to the RIGHT of the box of your choice, thus:

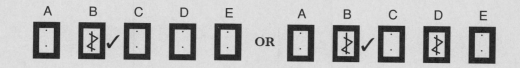

SECTION A

Answer questions 1–20 on the answer sheet.

1. The following graph shows how the displacement of an object varies with time.

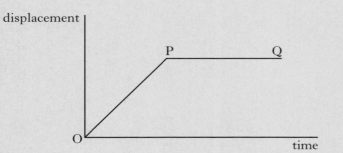

Which row of the table below best describes the motion of this object?

	From O to P	From P to Q
A	constant acceleration	constant velocity
B	zero velocity	constant deceleration
C	constant velocity	zero velocity
D	zero velocity	constant velocity
E	constant velocity	constant deceleration

2. Which of the following velocity-time graphs best describes a ball being thrown vertically into the air and returning to the thrower's hand?

A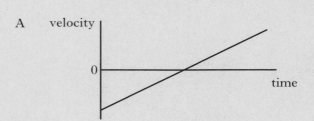

B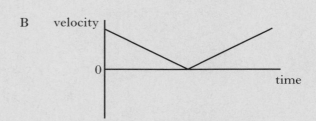

C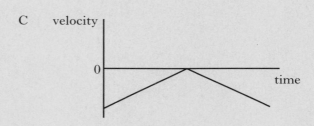

D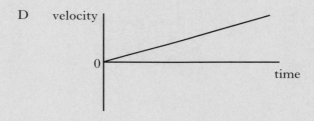

E

3. A force of 180 N is applied vertically upwards to a box of mass 15 kg.

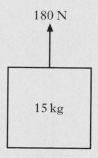

Assuming that the acceleration due to gravity is $9.8\,\text{m s}^{-2}$, the acceleration of the box is

A $2.2\,\text{m s}^{-2}$

B $7.6\,\text{m s}^{-2}$

C $9.8\,\text{m s}^{-2}$

D $12.0\,\text{m s}^{-2}$

E $19.6\,\text{m s}^{-2}$.

4. A box of mass 10 kg rests on an inclined plane. The component of the weight of the box acting down the incline is 50 N. A force of 300 N, parallel to the plane, is applied to the box as shown.

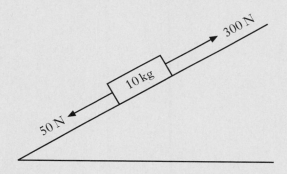

The box accelerates at $10\,\text{m s}^{-2}$ up the plane.

The size of the force of friction opposing the motion of the box is

A 50 N

B 100 N

C 150 N

D 200 N

E 250 N.

5. A flat bottomed test-tube containing aluminium rivets is floated in liquid A.

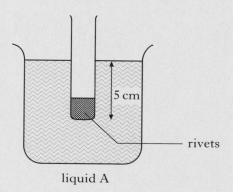

The bottom of the test-tube is at a depth of 5 cm below the surface.

The same test-tube and aluminium rivets are then floated in liquid B.

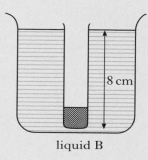

The bottom of the test-tube is at a depth of 8 cm below the surface.

Which of the following statement(s) is/are true?

I In each liquid the pressure at the bottom of the test-tube is the same.

II The density of liquid A is greater than the density of liquid B.

III In each liquid the upthrust on the bottom of the test-tube is the same.

A I only

B II only

C I and II only

D II and III only

E I, II and III

[Turn over

6. A helium filled balloon of mass 1·5 kg floats at a constant height of 100 m. The acceleration due to gravity is 9·8 m s^{-2}.

The upthrust on the balloon is

A   0 N

B   1·5 N

C   14·7 N

D   150 N

E   1470 N.

7. A sealed hollow buoy drifts from warm Atlantic waters into colder Arctic waters.

The volume of the buoy remains constant.

The pressure of the air trapped inside the buoy changes.

This is because the pressure of the trapped air is

A   directly proportional to the kelvin temperature

B   inversely proportional to the kelvin temperature

C   inversely proportional to the volume of the air in the buoy

D   inversely proportional to the celsius temperature

E   directly proportional to the celsius temperature.

8. In the following circuit the current from the battery is 3 A.

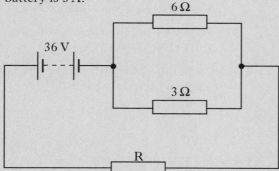

Assuming that the battery has negligible internal resistance, the resistance of resistor R is

A   3 Ω

B   4 Ω

C   10 Ω

D   12 Ω

E   18 Ω.

9. The diagram below shows a balanced Wheatstone bridge where all the resistors have different values.

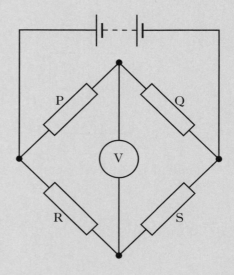

Which change(s) would make the bridge unbalanced?

I   Interchange resistors P and S.

II   Interchange resistors P and Q.

III   Change the e.m.f. of the battery.

A   I only

B   II only

C   III only

D   II and III only

E   I and III only

10. A student sets up the following circuit.

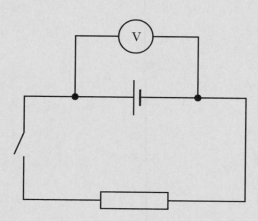

When the switch is open, the student notes that the reading on the voltmeter is 1·5 V. The switch is then closed and the reading falls to 1·3 V.

The decrease of 0·2 V is referred to as the

A e.m.f.

B lost volts

C peak voltage

D r.m.s. voltage

E terminal potential difference.

11. The unit for capacitance can be written as

A $V C^{-1}$

B $C V^{-1}$

C $J s^{-1}$

D $C J^{-1}$

E $J C^{-1}$.

12. Which of the following statements about capacitors is/are true?

  I Capacitors are used to block a.c. signals.

  II Capacitors are used to block d.c. signals.

  III Capacitors can store energy.

  IV Capacitors can store electric charge.

A I only

B I and III only

C II and III only

D II, III and IV only

E III and IV only

13. The operational amplifier connected in the circuit below is powered by a supply of +15 V and −15 V.

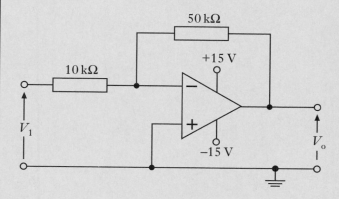

The input voltage V_1 is +5 V. The most likely value for the output voltage V_o is

A −25 V

B −13 V

C −1 V

D +13 V

E +25 V.

14. The amplifier shown below has an output voltage of 5·0 V.

Input voltage V_1 is originally 0·5 V and input voltage V_2 is originally 0·6 V.

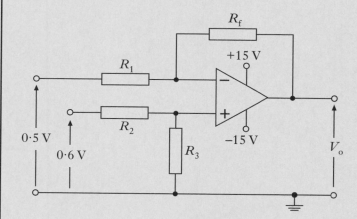

$R_1 = R_2$ and $R_f = R_3$

The input voltages V_1 and V_2 are increased to 1·0 V and 1·2 V respectively.

The output voltage V_o is now

A 0·2 V

B 2·2 V

C 5·0 V

D 6·0 V

E 10 V.

15. Microwave radiation is incident on a metal plate which has 2 slits, P and Q. A microwave receiver is moved from R to S, and detects a series of maxima and minima of intensity at the positions shown.

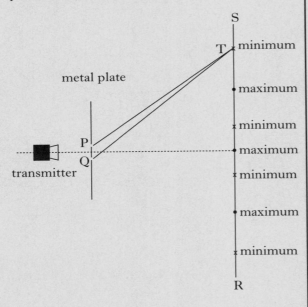

The microwave radiation has a wavelength of 4 cm.

The path difference between PT and QT is

A 2 cm

B 3 cm

C 4 cm

D 5 cm

E 6 cm.

16. Light of frequency 5.0×10^{14} Hz passes from air into a block of glass of refractive index 1·5.

Which row in the following table gives the correct values for the velocity, frequency and wavelength of the light in the glass?

	velocity/m s^{-1}	frequency/Hz	wavelength/m
A	2.0×10^8	5.0×10^{14}	4.0×10^{-7}
B	3.0×10^8	5.0×10^{14}	6.0×10^{-7}
C	3.0×10^8	3.3×10^{14}	6.0×10^{-7}
D	2.0×10^8	3.3×10^{14}	6.0×10^{-7}
E	3.0×10^8	3.3×10^{14}	4.0×10^{-7}

17. In a laser, a photon of radiation is emitted when an electron makes a transition from a higher energy level to a lower level, as shown below.

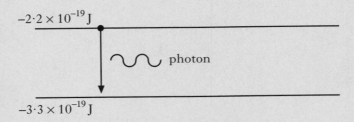

The energy in each pulse of radiation from the laser is 10 J. How many photons are there in each pulse?

A 1.8×10^{19}

B 3.0×10^{19}

C 3.7×10^{19}

D 4.5×10^{19}

E 9.1×10^{19}

18. In a darkened room, a small lamp is placed 2 cm from a photodiode which is connected in the circuit as shown. The lamp may be regarded as a point source. The reading on the ammeter is 27 µA.

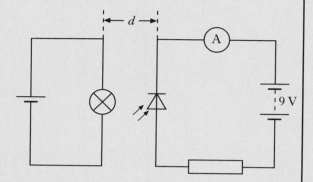

Which graph shows correctly how the ammeter reading changes as the distance d between the lamp and the photodiode is increased to 6 cm?

A
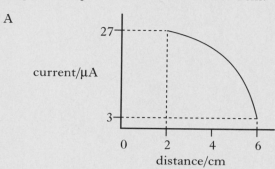

B
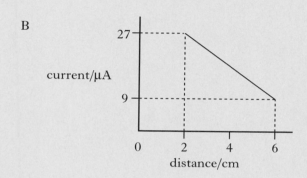

C
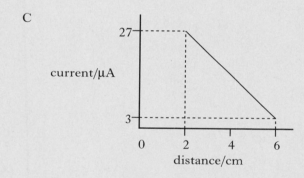

D
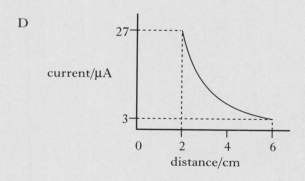

E
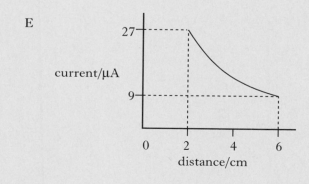

19. Which row of the table shows the correct values of x, y and z for the nuclear reaction described below?

$$^{214}_{x}\text{Pb} \rightarrow {}^{y}_{83}\text{Bi} + {}^{0}_{z}\text{e}$$

	x	y	z
A	84	214	1
B	83	210	4
C	85	214	2
D	82	214	−1
E	82	210	−1

20. The risk of biological harm from exposure to radiation depends on

I the absorbed dose

II the body organs exposed

III the type of radiation.

Which statement(s) is/are true?

A I only

B II only

C III only

D II and III only

E I, II and III

[SECTION B begins on *Page twelve*]

SECTION B

Write your answers to questions 21 to 30 in the answer book.

21. A basketball is held below a motion sensor. The basketball is released from rest and falls onto a wooden block. The motion sensor is connected to a computer so that graphs of the motion of the basketball can be displayed.

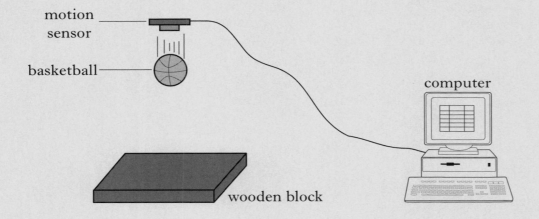

A displacement-time graph for the motion of the basketball from the instant of its release is shown.

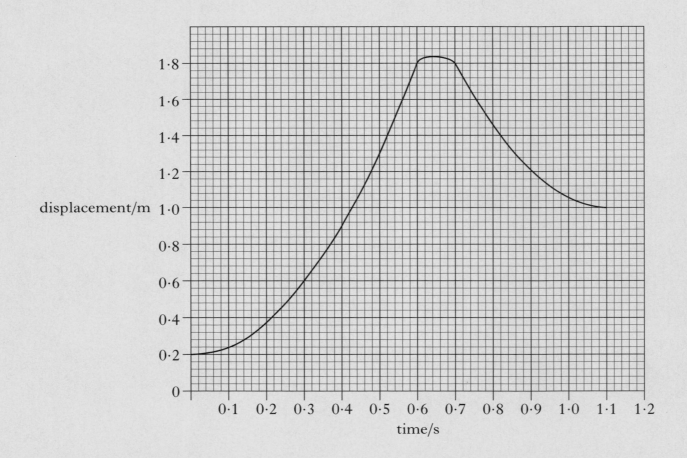

21. (continued)

(a) (i) What is the distance between the motion sensor and the top of the basketball when it is released?

(ii) How far does the basketball fall before it hits the wooden block?

(iii) Show, by calculation, that the acceleration of the basketball as it falls is $8.9\,\text{m s}^{-2}$.

3

(b) The basketball is now dropped several times from the same height. The following values are obtained for the acceleration of the basketball.

$8.9\,\text{m s}^{-2}$ $\quad$ $9.1\,\text{m s}^{-2}$ $\quad$ $8.4\,\text{m s}^{-2}$ $\quad$ $8.5\,\text{m s}^{-2}$ $\quad$ $9.0\,\text{m s}^{-2}$

Calculate:

(i) the mean of these values;

(ii) the approximate random uncertainty in the mean.

3

(c) The wooden block is replaced by a block of sponge of the same dimensions. The experiment is repeated and a new graph obtained.

Describe and explain any **two** differences between this graph and the original graph.

2

(8)

[Turn over

Marks

22. A technician designs the apparatus shown in the diagram to investigate the relationship between the temperature and pressure of a fixed mass of nitrogen which is kept at a constant volume.

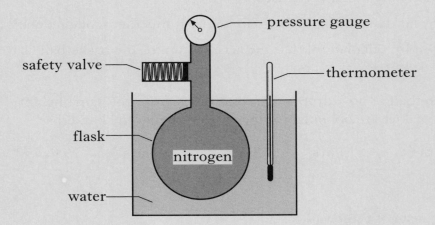

(a) The pressure of the nitrogen is 109 kPa when its temperature is 15 °C. The temperature of the nitrogen rises to 45 °C.

Calculate the new pressure of the nitrogen in the flask. **2**

(b) Explain, in terms of the movement of gas molecules, what happens to the pressure of the nitrogen as its temperature is increased. **2**

(c) The technician has fitted a safety valve to the apparatus.

A diagram of the valve is shown below.

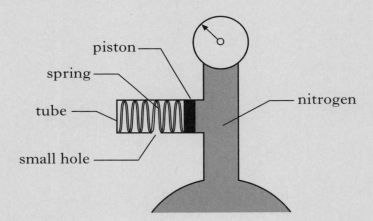

The piston of cross-sectional area $4.0 \times 10^{-6}\,\text{m}^2$ is attached to the spring. The piston is free to move along the tube.

The following graph shows how the length of the spring varies with the force exerted by the nitrogen on the piston.

22. (c) (continued)

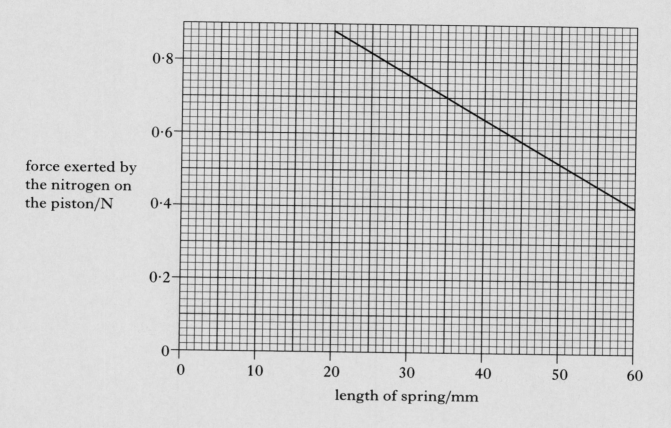

(i) Calculate the force exerted by the nitrogen on the piston when the reading on the pressure gauge is 1.75×10^5 Pa.

(ii) What is the length of the spring in the safety valve when the pressure of the nitrogen is 1.75×10^5 Pa?

3

(d) The technician decides to redesign the apparatus so that the bulb of the thermometer is placed inside the flask.

Give **one** reason why this improves the design of the apparatus.

1

(8)

[Turn over

Marks

23. (a) A space vehicle of mass 2500 kg is moving with a constant speed of $0.50\,\text{m s}^{-1}$ in the direction shown. It is about to dock with a space probe of mass 1500 kg which is moving with a constant speed in the opposite direction.

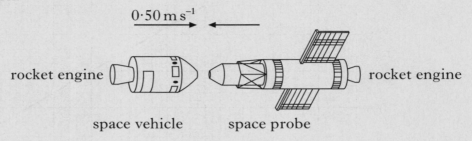

After docking, the space vehicle and space probe move off together at $0.20\,\text{m s}^{-1}$ in the original direction in which the space vehicle was moving.

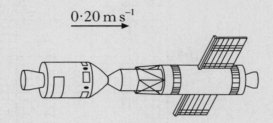

Calculate the speed of the space probe before it docked with the space vehicle. **2**

(b) The space vehicle has a rocket engine which produces a constant thrust of 1000 N. The space probe has a rocket engine which produces a constant thrust of 500 N.

The space vehicle and space probe are now brought to rest from their combined speed of $0.20\,\text{m s}^{-1}$.

(i) Which rocket engine was switched on to bring the vehicle and probe to rest?

(ii) Calculate the time for which this rocket engine was switched on. You may assume that a negligible mass of fuel was used during this time. **3**

(c) The space vehicle and space probe are to be moved from their stationary position at A and brought to rest at position B, as shown.

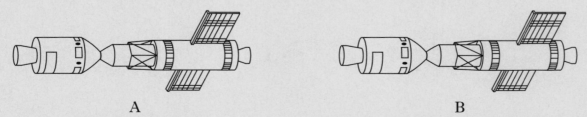

Explain clearly how the rocket engines of the space vehicle and the space probe are used to complete this manoeuvre.

Your explanation must include an indication of the relative time for which each rocket engine must be fired.

You may assume that a negligible mass of fuel is used during this manoeuvre. **2**

(7)

24. A battery has an e.m.f. of 6·0 V and internal resistance of 2·0 Ω.

(a) What is meant by an *e.m.f. of 6·0 V*? **1**

(b) The battery is connected in series with two resistors, R_1 and R_2. Resistor R_1 has a resistance of 20 Ω.

The reading on the ammeter is 200 mA.

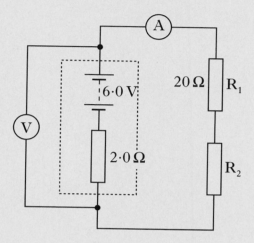

(i) Show by calculation that R_2 has a resistance of 8·0 Ω.

(ii) Calculate the reading on the voltmeter. **4**

(c) The battery is now connected to two identical lamps as shown below.

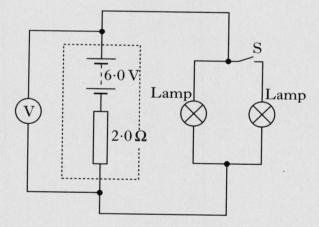

Describe and explain what happens to the reading on the voltmeter when switch S is closed. **2**

(7)

[Turn over

25. (a) The circuit below is used to investigate the charging of a $2000\,\mu F$ capacitor. The d.c. supply has negligible internal resistance.

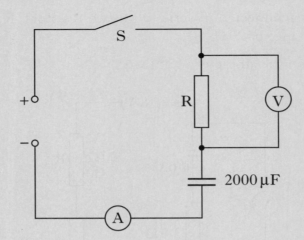

The graphs below show how the potential difference V_R across the **resistor** and the current I in the circuit vary with time from the instant switch S is closed.

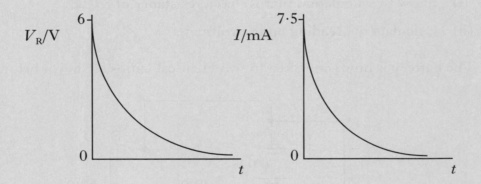

(i) What is the potential difference across the capacitor when it is fully charged?

(ii) Calculate the energy stored in the capacitor when it is fully charged.

(iii) Calculate the resistance of R in the circuit above.

5

25. (continued)

(b) The circuit below is used to investigate the charging and discharging of a capacitor.

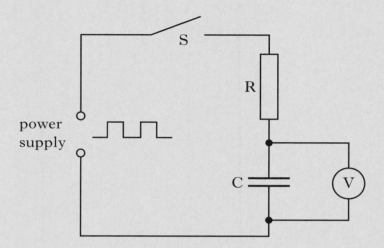

The graph below shows how the power supply voltage varies with time after switch S is closed.

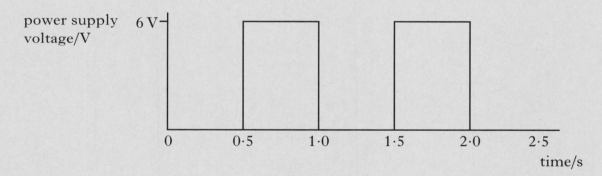

The capacitor is initially uncharged.

The capacitor charges fully in 0·3 s and discharges fully in 0·3 s.

Sketch a graph of the reading on the voltmeter for the first 2·5 s after switch S is closed.

The axes on your graph must have the same numerical values as those in the above graph.

2

(7)

[Turn over

Marks

26. An alternating voltage signal displayed on an oscilloscope screen is shown below. The peak voltage is 6·0 V and the time base setting is 2 ms/cm.

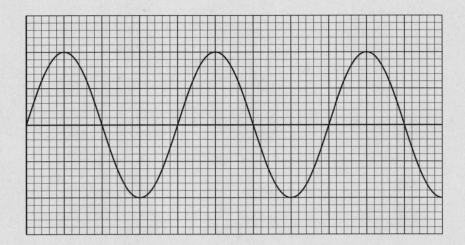

(a) Calculate the frequency of the signal. **2**

(b) This alternating voltage is used as the input voltage V_1 for the operational amplifier circuit shown below. R_f is a variable resistor.

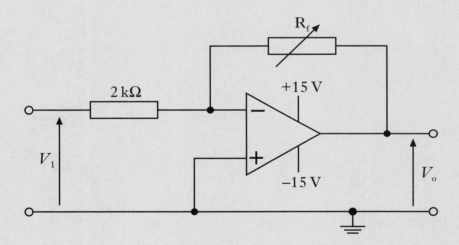

(i) In what mode is the operational amplifier operating?

(ii) The variable resistor R_f is set at 3·0 kΩ.

(A) On square ruled paper, sketch a graph of the output voltage V_o. Numerical values must be shown.

(B) Calculate the **r.m.s.** value of the output voltage V_o.

(iii) The resistance of resistor R_f is gradually increased from 3 kΩ to 8 kΩ. Describe what happens to the output voltage V_o during this time. **7**

(9)

27. A ray of red light is directed at a glass prism of side 80 mm as shown in the diagram below.

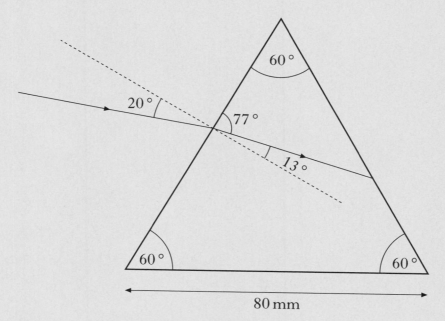

(a) Using information from this diagram, show that the refractive index of the glass for this red light is 1·52. **1**

(b) What is meant by the term *critical angle*? **1**

(c) Calculate the critical angle for the red light in the prism. **2**

(d) Sketch a diagram showing the path of the ray of red light until after it leaves the prism. Mark on your diagram the values of all relevant angles. **3**

(7)

[Turn over

28. An image intensifier is used to improve night vision. It does this by amplifying the light from an object.

Light incident on a photocathode causes the emission of photoelectrons. These electrons are accelerated by an electric field and strike a phosphorescent screen causing it to emit light. This emitted light is of a greater intensity than the light that was incident on the photocathode.

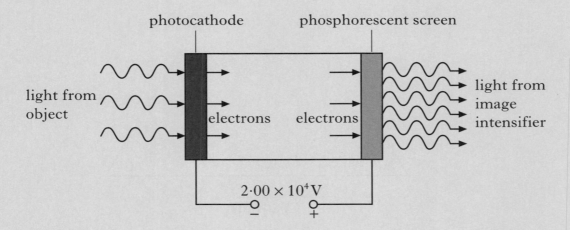

The voltage between the photocathode and the phosphorescent screen is $2 \cdot 00 \times 10^4$ V.

The minimum frequency of the incident light that allows photoemission to take place is $3 \cdot 33 \times 10^{14}$ Hz.

(a) What name is given to the minimum frequency of the light required for photoemission to take place?

(b) (i) Show that the work function of the photocathode material is $2 \cdot 21 \times 10^{-19}$ J.

(ii) Light of frequency $5 \cdot 66 \times 10^{14}$ Hz is incident on the photocathode. Calculate the maximum kinetic energy of an electron emitted from the photocathode.

(iii) Calculate the kinetic energy gained by an electron as it is accelerated from the photocathode to the phosphorescent screen.

29. (a) A sample of pure semiconducting material is doped by adding impurity atoms.

How does this addition affect the resistance of the semiconducting material? **1**

(b) The circuit below shows a p-n junction diode used as a light emitting diode (LED).

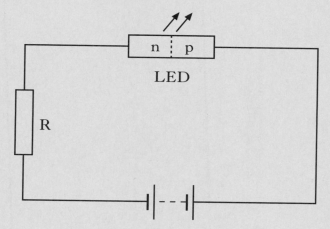

(i) Explain in terms of the charge carriers how the LED emits light.

(ii) Monochromatic light from the LED is incident on a grating as shown.

The spacing between lines in the grating is 5.0×10^{-6} m.

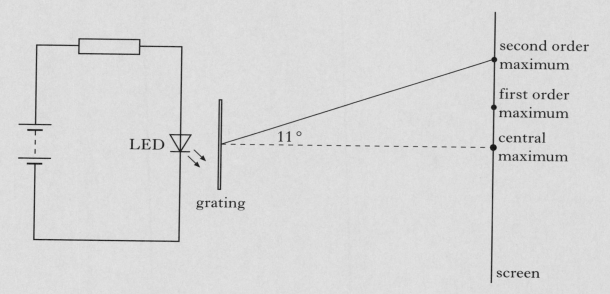

What is the wavelength of the light emitted by the LED? **4**

(5)

[Turn over for Question 30 on *Page twenty-four*

Marks

30. (a) Torbernite is a mineral which contains uranium.

The activity of 1·0 kg of pure torbernite is $5·9 \times 10^6$ decays per second.

A sample of material of mass 0·6 kg contains 40% torbernite. The remaining 60% of the material is not radioactive.

What is the activity of the sample in becquerels? **2**

(b) The table below gives the quality factor for some types of radiation.

Type of radiation	Quality factor
Gamma rays	1
Fast neutrons	10
Alpha particles	20

Exposure to 150 µGy of alpha particles for 6 hours gives the same dose equivalent rate as exposure for 8 hours to 400 µGy of one of the other radiations in the table above.

Identify this radiation.

You must justify your answer by calculation. **3**

(5)

[END OF QUESTION PAPER]

2003 HIGHER

X069/301

NATIONAL
QUALIFICATIONS
2003

MONDAY, 19 MAY
1.00 PM – 3.30 PM

**PHYSICS
HIGHER**

Read Carefully

1 All questions should be attempted.

Section A (questions 1 to 20)

2 Check that the answer sheet is for Physics Higher (Section A).
3 Answer the questions numbered 1 to 20 on the answer sheet provided.
4 Fill in the details required on the answer sheet.
5 Rough working, if required, should be done only on this question paper, or on the first two pages of the answer book provided—**not** on the answer sheet.
6 For each of the questions 1 to 20 there is only **one** correct answer and each is worth 1 mark.
7 Instructions as to how to record your answers to questions 1–20 are given on page three.

Section B (questions 21 to 29)

8 Answer questions numbered 21 to 29 in the answer book provided.
9 Fill in the details on the front of the answer book.
10 Enter the question number clearly in the margin of the answer book beside each of your answers to questions 21 to 29.
11 Care should be taken to give an appropriate number of significant figures in the final answers to calculations.

DATA SHEET
COMMON PHYSICAL QUANTITIES

Quantity	Symbol	Value	Quantity	Symbol	Value
Speed of light in vacuum	c	3.00×10^8 m s^{-1}	Mass of electron	m_e	9.11×10^{-31} kg
Magnitude of the charge on an electron	e	1.60×10^{-19} C	Mass of neutron	m_n	1.675×10^{-27} kg
Gravitational acceleration on Earth	g	9.8 m s^{-2}	Mass of proton	m_p	1.673×10^{-27} kg
Planck's constant	h	6.63×10^{-34} J s			

REFRACTIVE INDICES
The refractive indices refer to sodium light of wavelength 589 nm and to substances at a temperature of 273 K.

Substance	Refractive index	Substance	Refractive index
Diamond	2.42	Water	1.33
Crown glass	1.50	Air	1.00

SPECTRAL LINES

Element	Wavelength/nm	Colour	Element	Wavelength/nm	Colour
Hydrogen	656	Red	Cadmium	644	Red
	486	Blue-green		509	Green
	434	Blue-violet		480	Blue
	410	Violet		Lasers	
	397	Ultraviolet	Element	Wavelength/nm	Colour
	389	Ultraviolet	Carbon dioxide	9550 } 10590 }	Infrared
Sodium	589	Yellow	Helium-neon	633	Red

PROPERTIES OF SELECTED MATERIALS

Substance	Density/ kg m^{-3}	Melting Point/ K	Boiling Point/ K
Aluminium	2.70×10^3	933	2623
Copper	8.96×10^3	1357	2853
Ice	9.20×10^2	273	....
Sea Water	1.02×10^3	264	377
Water	1.00×10^3	273	373
Air	1.29	....	....
Hydrogen	9.0×10^{-2}	14	20

The gas densities refer to a temperature of 273 K and a pressure of 1.01×10^5 Pa.

SECTION A

For questions 1 to 20 in this section of the paper, an answer is recorded on the answer sheet by indicating the choice A, B, C, D or E by a stroke made in ink in the appropriate box of the answer sheet—see the example below.

EXAMPLE

The energy unit measured by the electricity meter in your home is the

    A ampere

    B kilowatt-hour

    C watt

    D coulomb

    E volt.

The correct answer to the question is B—kilowatt-hour. Record your answer by drawing a heavy vertical line joining the two dots in the appropriate box on your answer sheet in the column of boxes headed B. The entry on your answer sheet would now look like this:

If after you have recorded your answer you decide that you have made an error and wish to make a change, you should cancel the original answer and put a vertical stroke in the box you now consider to be correct. Thus, if you want to change an answer D to an answer B, your answer sheet would look like this:

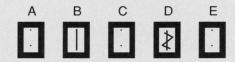

If you want to change back to an answer which has already been scored out, you should enter a tick (✓) to the RIGHT of the box of your choice, thus:

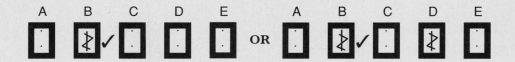

SECTION A

Answer questions 1–20 on the answer sheet.

1. Which of the following are **both** vectors?

   A Speed and weight

   B Kinetic energy and potential energy

   C Mass and momentum

   D Weight and momentum

   E Force and speed

2. A vehicle is travelling in a straight line.

   Graphs of velocity and acceleration against time are shown below.

   Which pair of graphs could represent the motion of the vehicle?

   A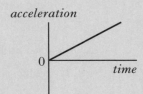

   B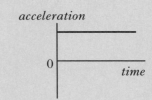

   C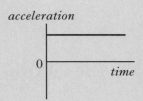

   D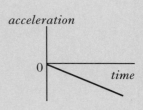

   E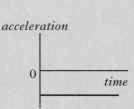

3. A block of mass 4·0 kg and a block of mass 6·0 kg are linked by a spring balance of negligible mass.

The blocks are placed on a frictionless horizontal surface. A force of 18·0 N is applied to the 6·0 kg block as shown.

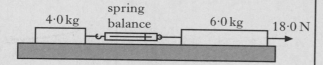

What is the reading on the spring balance?

A 7·2 N
B 9·0 N
C 10·8 N
D 18·0 N
E 40·0 N

4. A car of mass 1000 kg is travelling at a speed of 40 m s^{-1} along a straight road. The brakes are applied and the car decelerates to 10 m s^{-1}.

How much kinetic energy is lost by the car?

A 15 kJ
B 50 kJ
C 450 kJ
D 750 kJ
E 800 kJ

5. A car is designed with a "crumple-zone" so that the front of the car collapses during impact.

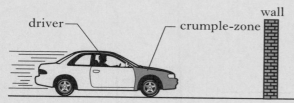

The purpose of the crumple-zone is to

A decrease the driver's change in momentum per second

B increase the driver's change in momentum per second

C decrease the driver's final velocity

D increase the driver's total change in momentum

E decrease the driver's total change in momentum.

6. A fixed mass of gas condenses at atmospheric pressure to form a liquid.

Which row in the table shows the approximate increase in density and the approximate decrease in spacing between molecules?

	Approximate increase in density	Approximate decrease in spacing between molecules
A	10 times	2 times
B	100 times	10 times
C	1000 times	10 times
D	1 000 000 times	100 times
E	1 000 000 times	1000 times

[Turn over

7. A rigid metal cylinder stores compressed gas. Gas is gradually released from the cylinder. The temperature of the gas remains constant.

Which set of graphs shows how the pressure, the volume and the mass of the gas **in the cylinder** change with time?

A
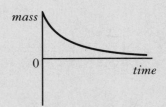

B
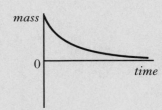

C
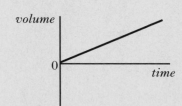 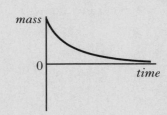

D
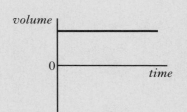 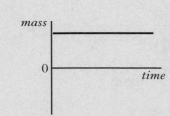

E
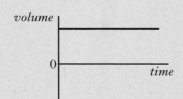 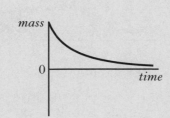

8. Two parallel metal plates, R and S, are connected to a 2·0 V d.c. supply as shown.

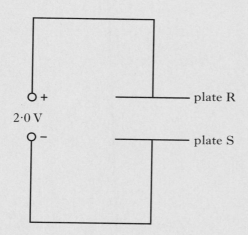

An electron is moved from plate R to plate S.

The gain in electrical potential energy of the electron is

A $8·0 \times 10^{-20}$ J

B $1·6 \times 10^{-19}$ J

C $3·2 \times 10^{-19}$ J

D $6·4 \times 10^{-19}$ J

E $1·3 \times 10^{-19}$ J.

9. In the following circuit, the battery has an e.m.f. of 8·0 V and an internal resistance of 0·20 Ω.

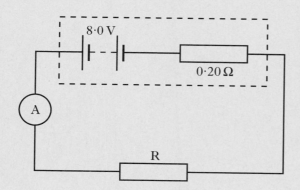

The reading on the ammeter is 4·0 A.

The resistance of R is

A 0·5 Ω

B 1·8 Ω

C 2·0 Ω

D 2·2 Ω

E 6·4 Ω.

10. In the following circuit, the supply has negligible internal resistance.

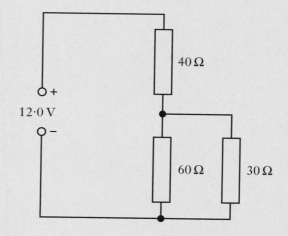

The p.d. across the 30 Ω resistor is

A 8·0 V

B 7·2 V

C 6·0 V

D 4·8 V

E 4·0 V.

11. A student sets up the following circuit.

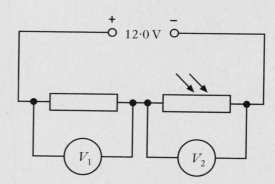

The intensity of light incident on the LDR is reduced.

Which row in the table shows the effect on the voltmeter readings V_1 and V_2?

	V_1	V_2
A	increases	increases
B	decreases	decreases
C	increases	decreases
D	decreases	increases
E	no change	increases

12. A student writes the following statements about a capacitor.

   I The current in a circuit containing a capacitor decreases when the supply frequency increases.

   II A capacitor can store charge.

   III A capacitor can block d.c.

Which of these is/are correct?

A I only

B II only

C III only

D I and II only

E II and III only

13. A farad is a

A volt per ampere

B volt per ohm

C coulomb per volt

D coulomb per second

E joule per coulomb.

14. A $10\,\mu\text{F}$ capacitor is connected to a $50\,\text{V}$ supply. The maximum charge stored by the capacitor is

A $2 \cdot 0 \times 10^{-7}\,\text{C}$

B $5 \cdot 0 \times 10^{-4}\,\text{C}$

C $5 \cdot 0\,\text{C}$

D $5 \cdot 0 \times 10^{2}\,\text{C}$

E $5 \cdot 0 \times 10^{6}\,\text{C}$.

15. In the following passage three words have been replaced by the letters **X**, **Y** and **Z**.

"*Monochromatic light is incident on a grating and the resulting interference pattern is viewed on a screen. The distance between neighbouring areas of constructive interference on the screen*:

is ....**X**........ *when the screen is moved further away from the grating*;

is ....**Y**........ *when light of a greater wavelength is used*;

is ....**Z**........ *when the distance between the slits is increased*."

Which row of the table shows the missing words?

	X	Y	Z
A	increased	increased	increased
B	increased	increased	decreased
C	decreased	decreased	increased
D	decreased	decreased	decreased
E	increased	decreased	decreased

16. An engineer creates an experimental window using sheets of transparent plastics **P**, **Q** and **R**.

    A ray of light directed at the window follows the path shown.

    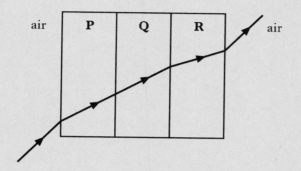

    Which row in the table gives possible values for the refractive indices of the three plastics?

	P	Q	R
A	1·5	1·9	2·3
B	1·5	1·5	2·3
C	2·3	2·3	1·5
D	2·3	1·9	1·5
E	1·5	1·5	1·2

17. A unit for the intensity of light is

    A $J m^{-1}$

    B $J m^{-2}$

    C $J s^{-1} m^{-1}$

    D $J s^{-1} m^{-2}$

    E $J s^{-2} m^{-2}$.

18. When light of frequency f is shone on to a certain metal, photoelectrons are ejected with a maximum velocity v and kinetic energy E_k.

    When light of the same frequency and twice the intensity is shone on the same surface then

    I twice as many electrons are ejected per second

    II the speed of the fastest electrons is now $2v$

    III the kinetic energy of the fastest electrons is now $2E_k$.

    Which of the statements above is/are correct?

    A I only

    B II only

    C III only

    D I and II only

    E II and III only

19. A student writes the following statements about n-type semiconductor material.

    I Most charge carriers are negative.

    II The n-type material has a negative charge.

    III Impurity atoms in the material have 5 outer electrons.

    Which of these statements is/are true?

    A I only

    B II only

    C III only

    D I and II only

    E I and III only

20. Which of the following statements describes nuclear fission?

    A A nucleus of large mass number splits into two nuclei, releasing several neutrons.

    B A nucleus of large mass number splits into two nuclei, releasing several electrons.

    C A nucleus of large mass number splits into two nuclei, releasing several protons.

    D Two nuclei combine to form one nucleus, releasing several electrons.

    E Two nuclei combine to form one nucleus, releasing several neutrons.

[SECTION B begins on *Page eleven*]

SECTION B

Write your answers to questions 21 to 29 in the answer book.

21. A golfer on an elevated tee hits a golf ball with an initial velocity of 35·0 m s^{-1} at an angle of 40° to the horizontal.

    The ball travels through the air and hits the ground at point R.

    Point R is 12 m below the height of the tee, as shown.

    diagram not to scale

    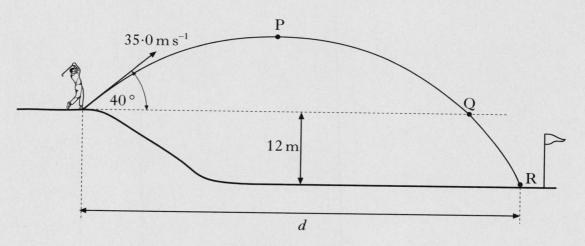

    The effects of air resistance can be ignored.

    (a) Calculate:

    (i) the horizontal component of the initial velocity of the ball;

    (ii) the vertical component of the initial velocity of the ball;

    (iii) the time taken for the ball to reach its maximum height at point P. **4**

    (b) From its maximum height at point P, the ball falls to point Q, which is at the same height as the tee.

    It then takes a further 0·48 s to travel from Q until it hits the ground at R.

    Calculate the total horizontal distance *d* travelled by the ball. **3**

    (7)

    [Turn over

22. Two ice skaters are initially skating together, each with a velocity of $2\cdot2\,\text{m s}^{-1}$ to the right as shown.

The mass of skater R is $54\,\text{kg}$. The mass of skater S is $38\,\text{kg}$.

Skater R now pushes skater S with an average force of $130\,\text{N}$ for a short time. This force is in the same direction as their original velocity.

As a result, the velocity of skater S increases to $4\cdot6\,\text{m s}^{-1}$ to the right.

(a) Calculate the magnitude of the change in momentum of skater S.   2

(b) How long does skater R exert the force on skater S?   2

(c) Calculate the velocity of skater R immediately after pushing skater S.   2

(d) Is this interaction between the skaters elastic?

You must justify your answer by calculation.   3

(9)

Marks

23. A tank of water rests on a smooth horizontal surface.

    (a) A student takes measurements of the pressure at various depths below the surface of the water, using the apparatus shown.

    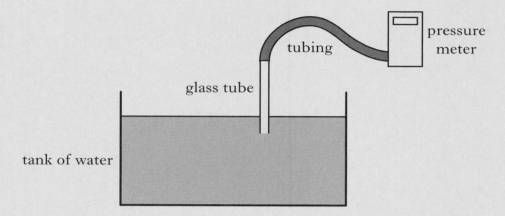

    The pressure meter is set to zero before the glass tube is lowered into the water.

    (i) Sketch a graph to show how the pressure due to the water varies with depth below the surface of the water.

    (ii) Calculate the pressure due to the water at a depth of 0·25 m below its surface.

    (iii) As the glass tube is lowered further into the tank, the student notices that some water rises inside the glass tube. Explain why this happens. **4**

    (b) The mass of water in the tank is $2·7 \times 10^3$ kg. The tank has a mass of 300 kg and a flat rectangular base. The dimensions of the tank are shown in the diagram below.

    Atmospheric pressure is $1·01 \times 10^5$ Pa.

    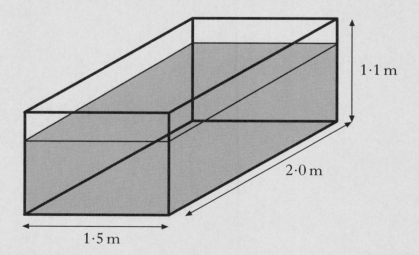

    Calculate the total pressure exerted by the base of the tank on the surface on which it rests. **3**

    (7)

Marks

24. A technician designs the following apparatus to investigate the pressure of a gas at different temperatures.

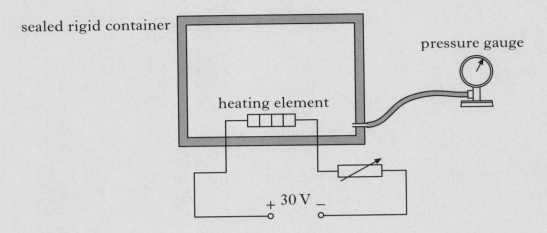

The heating element is used to raise the temperature of the gas.

(a) Initially the gas is at a pressure of 1.56×10^5 Pa and a temperature of 27 °C. The temperature of the gas is then raised by 50 °C.

Calculate the new pressure of the gas in the container. **2**

(b) The power supply shown above has an e.m.f. of 30 V and negligible internal resistance. The resistance of the heating element is 0.50 Ω and the resistance of the variable resistor is set to 1.50 Ω.

(i) Calculate the power output from the heating element.

(ii) How would your answer to part (b)(i) be affected if the internal resistance of the power supply was **not** negligible? You must justify your answer. **4**

(6)

25. (a) A signal generator is connected to an oscilloscope. The output of the signal generator is set to a peak voltage of 15 V.

The following diagram shows the trace obtained, the Y-gain and the timebase controls on the oscilloscope. The scale for the Y-gain has been omitted.

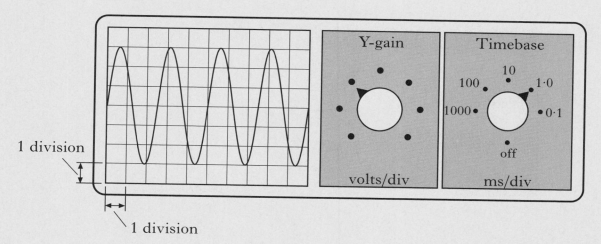

Calculate:

(i) the Y-gain setting of the oscilloscope;

(ii) the frequency of the signal in hertz.   **3**

(b) The signal generator is now connected in the circuit shown below.

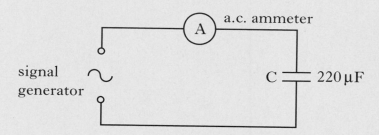

The signal generator is adjusted to give a peak output voltage of 12 V at a frequency of 300 Hz. The internal resistance of the signal generator and the resistance of the a.c. ammeter are negligible.

(i) Calculate the r.m.s. value of the output voltage from the signal generator.

(ii) Calculate the **maximum** energy stored by the capacitor during one cycle of the supply voltage.

(iii) The frequency of the signal generator is gradually increased.
What happens to the reading on the ammeter?

(iv) When a capacitor is connected to a d.c. supply, the current quickly falls to zero. Explain why the current does **not** fall to zero in the circuit above.   **6**

(9)

[Turn over

26. A washing machine is filled with water, emptied and refilled several times during a wash cycle. A water level detector is used to ensure the water does not overflow.

One design of water level detector uses a specially shaped glass prism, as shown below.

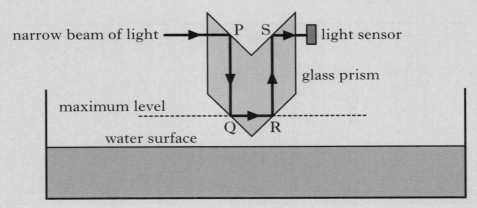

When the water in the machine is below the maximum level indicated in the diagram, the light sensor is illuminated by the narrow beam of light.

(a) The light sensor consists of an LDR connected in a Wheatstone bridge circuit with values of resistance as shown.

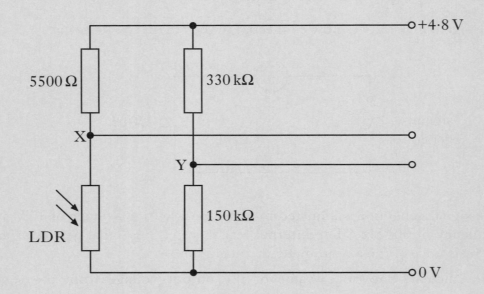

When the water in the machine is at the maximum level, the bridge is balanced.

Calculate the resistance of the LDR when the bridge is balanced.

26. (continued)

(b) Points X and Y of the Wheatstone bridge are connected to the inputs of an op-amp circuit as shown.

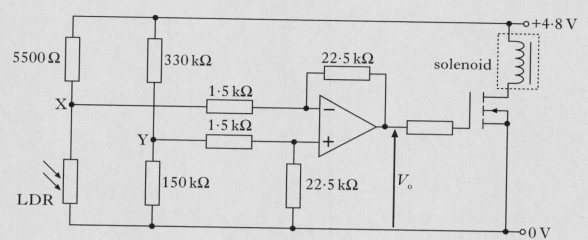

The potential at Y is 1·50 V. When the washing machine is filling with water, the narrow beam of light illuminates the LDR, the bridge is unbalanced and the potential at X is 1·28 V.

(i) Name the component in the circuit which has the following symbol.

(ii) Calculate the output voltage V_o of the op-amp when the LDR is illuminated.

(iii) When there is a current in the solenoid, it holds a valve open and water flows into the washing machine.

When the water reaches the maximum level, the valve closes.

Explain how the circuit causes the valve to close when the water reaches the maximum level.

(c) When the water is at the maximum level, the narrow beam of light no longer illuminates the light sensor, because light leaves the prism at Q.

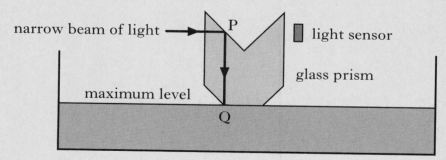

Explain why the light leaves the prism at Q.

27. (a) Electrons which orbit the nucleus of an atom can be considered as occupying discrete energy levels.

The following diagram shows some of the energy levels for a particular atom.

E_4 ———————— -1.4×10^{-19} J
E_3 ———————— -2.4×10^{-19} J

E_2 ———————— -5.6×10^{-19} J

E_1 ———————— -21.8×10^{-19} J

(i) The transition between which two of these energy levels produces radiation with the longest wavelength? You must justify your answer.

(ii) Calculate the frequency of the photon produced when an electron falls from E_3 to E_2.

5

(b) A laser produces light of frequency 4.74×10^{14} Hz in air.

A ray of light from this laser is directed into a block of glass as shown below.

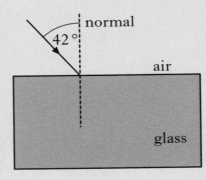

The refractive index of the glass for this light is 1.60.

(i) What is the value of the frequency of the light in the block of glass?

(ii) Calculate the wavelength of the light in the glass.

4

(9)

28. (a) An experiment with microwaves is set up as shown below.

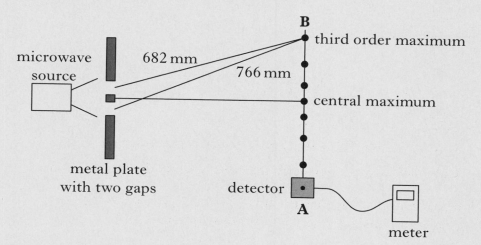

(i) As the detector is moved from **A** to **B**, the reading on the meter increases and decreases several times.

Explain, in terms of waves, how the pattern of maxima and minima is produced.

(ii) The measurements of the distance from each gap to a third order maximum are shown. Calculate the wavelength of the microwaves.

(b) A microphone is placed inside the cockpit of a jet aircraft.

The microphone is connected to the input terminals of the op-amp circuit shown below.

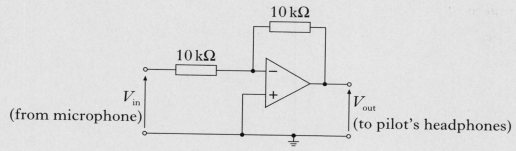

A noise in the cockpit produces the following signal from the microphone.

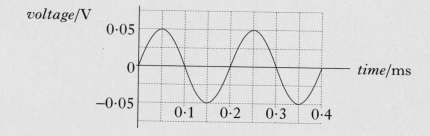

(i) Sketch a graph of the corresponding output voltage V_{out} against time.

Values are required on both axes.

(ii) The output from the op-amp is connected to the pilot's headphones.

Explain why the sound produced by the headphones **reduces** the noise level heard by the pilot.

29. A technician is studying samples of radioactive substances.

(a) The following statement describes a nuclear decay in one of the samples used by the technician.

$$^{238}_{92}U \rightarrow {}^{234}_{90}Th + {}^{4}_{2}He$$

(i) What type of particle is emitted during this decay?

(ii) In this sample $7\cdot2 \times 10^5$ nuclei decay in two minutes.

Calculate the average activity of the sample during this time.

3

(b) The technician now studies the absorption of the radiation emitted from a different sample using the apparatus shown below.

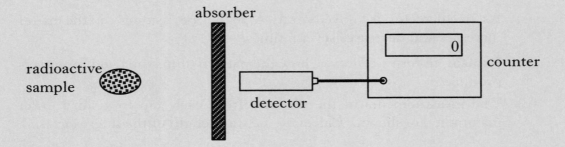

Different thicknesses of the absorber are placed in turn between the sample and the detector. For each thickness, the technician makes **repeated** measurements to obtain an average corrected count rate.

These results are then used to produce the following graph.

29. (b) (continued)

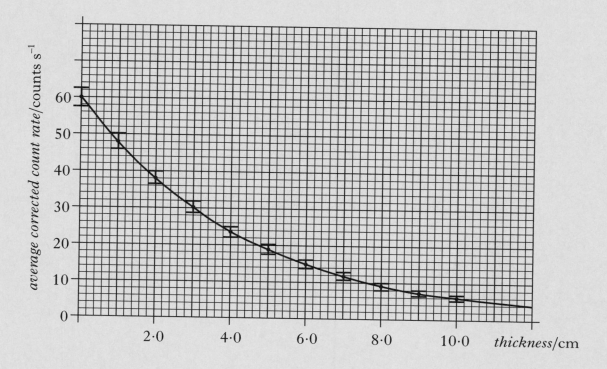

(i) Use the graph to calculate the half-value thickness of the absorber material for this radiation.

(ii) The technician has plotted each value of the average corrected count rate as a point with a vertical "bar" as shown.

Suggest a reason for this.

3

(c) The technician receives a total dose equivalent of 6.4×10^{-5} Sv from these two sources.

The quality factor of the radiation used in part (a) is 20.

The absorbed dose received by the technician from the source used in part (b) is 1.2×10^{-5} Gy. The quality factor of this radiation is 1.

Calculate the absorbed dose received by the technician from the source used in part (a).

2

(8)

[END OF QUESTION PAPER]

[BLANK PAGE]

[BLANK PAGE]

[BLANK PAGE]